P9-EMH-308

BODIES of EVIDENCE

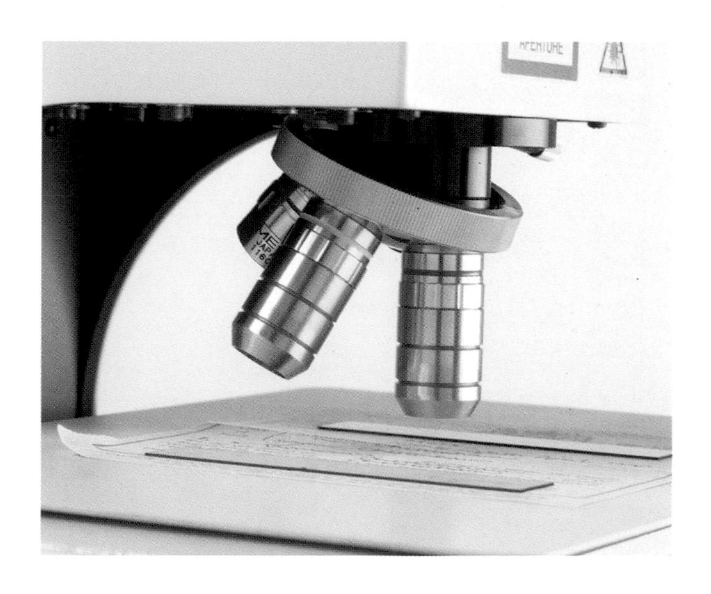

BODIES of EVIDENCE

Forensic Science and Crime

First Lyons Press edition, 2006

The Lyons Press is an imprint of The Globe Pequot Press.

10 9 8 7 6 5 4 3 2 1

Printed in China

Credits
Editor: Shaun Barrington
Designer: Cara Rogers
Production: Kate Rogers
Reproduction: Anorax Imaging Ltd

ISBN13 978-1-59228-580-8
ISBN 1-59228-580-5
Library of Congress Cataloging-in-Publication Data is available on file.

Contents

Foreword

By Dr. Lowell J. Levine, D.D.S., Director of the Medicolegal Investigations Unit of the New York State Police.

The technology employed by the forensic sciences in the 21st century represents a quantum leap from 1968, when I began my career in the forensic sciences. The advances in all the traditional fields of forensic sciences as well as the addition of areas that didn't exist almost forty years ago, such as DNA and Electronic Crime, have completely changed all aspects of crime scene processing, and all for the better.

Forensics is not always small-scale and in the lab. The partially reassembled wreckage of TWA Flight 800 stands in a National Transportation Safety Board hangar in Calverton, NY, July 16, 2001, five years after the crash. Investigators stated, after an astonishing piece of reconstruction, that the 747 came down off Long Island from a fire in the center fuel tank. As is often the case with public, or federal investigations into large-scale loss of life (in this case all 230 aboard), many have questioned the findings.

Evidence harvested from the crime scene, when analyzed by the scientist, can yield vital information that allows the crime to be solved. The recognition, documentation, proper collection, proper storage, and submission to a competent laboratory constitutes the most crucial part of the entire forensic scientific process. Obviously, unless the crime scene or evidence technician captures potential evidence, it will never become useful in the solution of the crime. His or her education and training has become the most important link in the forensic sciences chain. In the early days of my career, fingerprints accidentally left at a crime scene were probably the major contaminant brought to the scene by law enforcement and medical examiner/coroner personnel. Usually, a fingerprint check was part of the employment process so that law enforcement and medical examiner/coroner personnel's fingerprints could be compared and excluded from the "unknown" fingerprints found at a crime scene.

An anthropologist working on behalf of human rights carefully uncovers the remains of one of the "disappeared" in Argentina during the 1980s. Initially known only by number, many of the murder victims would later be identified based on their teeth and other forensic evidence. Such work has become ever more common, from natural disasters to terrorist attacks.

Today, testing for DNA is extremely sensitive, and biological matter can be a major contaminant that is brought to the crime scene by law enforcement and medical examiner/coroner personnel, along with other types of evidence accidentally carried into a crime scene. The processing of the crime scene in the O.J. Simpson case highlighted the problem of inadequate and sloppy crime scene discipline. That case, more than any scientific effort in my opinion, has caused a significant change in the manner in which crime scenes are now being processed.

Personnel going into a scene now resemble surgeons going into an operating room. Only essential personnel are allowed to go into the scene and others, such as supervisors, view the crime scene via very high-resolution video and still, electronically captured images. DNA sequence data of scientists or officers who worked the crime scene evidence is increasingly being kept in a database so their accidental contamination of a scene can be resolved.

In the foreseeable future, another issue that will need to be addressed by agencies doing crime scene work will involve the collection of DNA sequence data from employees: how it is databased, safeguards to prevent abuse, whether it is kept in perpetuity, as well as other issues.

As forensic science progresses, unsolved cases can be reexamined. The Oakland County (Michigan) Sheriff's Department forensic team search the edge of a backyard pool in Hampton Township with a metal detector in July 2003. Amazingly, they were looking (unsuccessfully) for a briefcase that an informant said held information about the disappearance of Teamster's Union boss Jimmy Hoffa 28 years earlier.

New types of digital evidence such as computer hard drives, cell phones, and other electronics didn't exist in 1968. They are now very important pieces of evidence that crime scene personnel are taught how to safeguard so that their potential as evidence may be maximized.

Today, all crime scene personnel are specially selected, highly trained individuals rather than, as sometimes used to be the case, junior officers, impaired officers, or officers with a disability that prevented them from performing normal duties. Significant amounts of time and money are invested in their training. As an example, all forensic investigative unit personnel of the New York State Police have received special uniform training in the location and recovery of human remains; the trainer was the Scientific Director and personnel of the JPAC CIL based at Hickam Air Force Base in Hawaii, the institution that recovers remains of United States war dead from all prior wars, and the leading facility of its type in the world. The training has included lectures, demonstrations, and practical experiences.

The American Society of Crime Lab Directors/LAB administers a program in Crime Scene Operations for their accredited institutions. Other organizations sponsor certification and accreditation programs for crime scene technicians and units.

Right now in the United States, the most popular television programs are the crime scene programs, which have grown from a beginning set in Las Vegas to a second set in Miami and a third set in New York City. They have supplanted the "Profiler" type programs in popularity. Today young people seeking forensic careers have migrated from the desire to be profilers to the desire to become crime scene technicians, based upon the glamorous image created by the media. In fact, crime scene personnel often spend long, tedious hours in difficult locations recovering evidence that is quite offensive to the senses.

The *CSI*-type programs have now caused a "CSI effect" among jurors and others who have very different expectations than the realities of crime scene processing can fulfill. Prosecutors often view this as a problem and they must therefore try very hard to expose jurors to the realities of the profession.

The *CSI* craze has also had some beneficial effects. The public no longer is impatient for the crime scene to be released by the authorities. The public also realizes that cutting-edge technology is expensive. The law enforcement community realizes the value of proper processing of the crime scene.

Telemedicine is being adapted to the problem as "Teleforensics." As the technology to send very accurate images of crime scenes digitally improves, experts far removed from the scene will be able to analyze portions of the scene. Although they may be great distances from the actual scene, their expertise will be available to help with the interpretation and gathering of evidence.

As long as the members of the forensic community are willing to be innovative, they will do better and better crime scene processing. The most difficult thing for the future is that the technology is changing so rapidly, and becoming so increasingly expensive that it's hard to budget and predict the future needs of the crime scene personnel.

Twenty years ago, before the dawn of the DNA revolution and the debacle of the O.J. Simpson case, Dr. Scott Christianson and his boss, Lawrence T. Kurlander, the New York State Director of Criminal Justice, were instrumental in establishing a world-class forensic unit and the Henry F. Williams Homicide Seminar that has since grown to become one of the premier annual training programs of its kind in the world. *Bodies of Evidence* recounts some of what has transpired in forensic science over the years.

Albany, N.Y., January 2006.

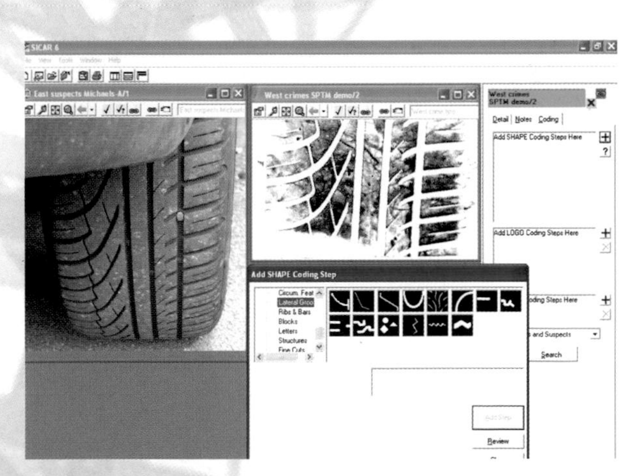

The first step in identifying tire marks—and the same goes for shoe prints—is not always a match of the mark to a specific car. It may be to identify the tire make (or shoe manufacturer). Computer databases can take the "legwork" out of such identification by cross-matching via elemental patterns of lines, waves, zigzags, diamonds, blocks, and circles.

Introduction:
Bodies of Evidence

***Corpus delicti*, meaning literally the "body of the crime," refers to the body of evidence that constitutes the offense; it is the objective proof or foundation and substance to establish that a crime has or has not been committed. And forensic science can be crucial to finding physical evidence to do this.**

Proving that a crime occurred and amassing evidence to prove how it was done and who did it can prove incredibly difficult and challenging. Clues found at the crime scene or obtained from an autopsy and laboratory analysis can provide police with a powerful tool. To make the most of it, investigators must combine what science has to offer with their own investigative experience and expertise, practicing close teamwork and a commitment to playing by the rules.

The truth is not always clear or readily available and investigators—be they police detectives or laboratory scientists—must keep an open mind and strive to gather convincing proof to support their conclusions. Otherwise they lose their credibility.

Florida Department of Law forensic team examine the body of a woman found by the Hamilton County Sheriff's Department. Securing the crime scene is the first priority. The victim's hands are bagged to preserve possible trace or DNA evidence.

Crime Scene Response

Prior to World War II, scouring a crime scene was likely to yield few definitive forensic clues.

In the 1930s, the primary police response to a crime scene was to obtain information about the appearance of the scene so that the witnesses, investigators, prosecutors, judges, and jury could get a clear sense of what happened. Cops were trained to take careful measurements and notes along with sketches and diagrams showing directions and scale. Detectives also relied on crime scene photographs taken by specially trained police photographers. They looked for fingerprints, palm prints, footprints, weapons, bullets, clothing, or other evidence left behind by the perpetrator, but depended most of all on witnesses, informers, and confessions.

As police became more proficient at using fingerprints to catch criminals, some offenders took to wearing gloves, and wiping guns and other incriminating objects. John Dillinger the famous gangster and some of his ilk even resorted to acid dips or plastic surgery to try and erase their telltale prints.

The police also placed special value in bloodstains, particularly if blood was found on a suspect's fingernails, clothes, or car. Sometimes blood spots were examined under an ultra-violet light or magnifying glass, but methods to determine the character of the stain were rather limited. The best that could be done was to identify a stain as probable human blood. If enough blood was available, laboratory technicians might attempt to identify it by blood group. Likewise, semen stains could sometimes be detected under ultra-violet light and fresh fluid semen could be examined under a microscope to yield a positive result indicating that it was human sperm.

Crime scene responders also looked for hair, but its utility was limited to identifying it by race, color, length, and determining whether it came from the head or some other part of the body. Nevertheless, some juries could be very impressed by the presence of a hair the prosecutor said belonged to the defendant, if it turned up at the scene of a crime.

Forensic scientists since Edmond Locard (1877–1966) of Lyons, France, recognized that *every contact leaves a trace*: "On the one hand, the criminal leaves marks at the crime scene of his passage; on the other hand, by inverse action, he takes with him, on his body or on his clothing, evidence of his stay or of his deed. Left or received, these traces are of extremely varied types." Locard's exchange principle emerged as one of the pillars of modern crime scene investigation, along with the power of logic and the rigorous application of the scientific method.

Detectives increasingly realized that some trace evidence could yield potential clues, particularly if it revealed a specific, unusual type of clothing, wood, or other object that was linked to a suspect. Time of death could be generally estimated based upon body temperature, rigor mortis, postmortem lividity, and degree of decomposition. However, homicide investigators and even medical examiners usually employed rough formulas to arrive at an estimated time of death.

By the mid-1930s, the Federal Bureau of Investigation and several major cities and states had begun to establish their own scientific crime laboratories, equipped with microscopes and other tools as well as chemical know-how, marshaled to harness the power of modern science in the war against crime.

In 1935 the most advanced textbook of its day, *Modern Criminal Investigation*, by Harry Soderman and John J. O'Connell—a work that was praised by Lewis J. Valentine, the police commissioner of the city of New York, as "an epoch-making contribution to the art and science of police procedure"—laid out an abbreviated list of what needed to be done by the first police detective upon his arrival at the scene of a homicide. The authors said he or she should:

- Ascertain who the perpetrator is and arrest immediately if possible.
- Note time of arrival.
- Expose shield and hold everyone at scene for questioning.
- Notify station or headquarters, giving a brief outline of case.
- Prevent anyone from touching body or disturbing anything, pending arrival of medical examiner or coroner, homicide squad, and technicians from police laboratory.
- Prevent unauthorized persons from entering upon the scene.
- Take names and addresses of all persons present and endeavor to ascertain name of perpetrator or perpetrators and a detailed identifiable description for immediate alarm.
- Prevent destruction of evidence such as fingerprints, footprints, etc.
- Clear room and immediate area of all but authorized persons present upon official business or detained on case.
- Keep witnesses separate to prevent conversation.
- Assign specific task to each detective—proceeding in a systematic manner. Keep record of all assignments and detail of work.

Regarding the examination of the crime scene itself, the authors prescribed the following:

- The position of the body is examined.
- The clothing and its position are noted.
- Traces on the body and on the clothing are noted, photographed, and sketched.
  (Only after the above facts have been ascertained, should the body be allowed to be moved and its position altered.)
- The wounds are examined by the medical examiner. The back of the body is examined, and the ground under it.
  (If something has been altered before the arrival of the homicide squad, the original position of objects sketched and photographed should be established with the aid of witnesses.)
- The weather (a) when the crime was discovered and
  (b) when the homicide squad arrives should be noted, especially in rural crimes.

(Position of sun and moon, rain, snow, frost, thaw, visibility, direction of wind, force of wind.)

- Examination of doors and windows, furniture, etc., will disclose the probable direction of entry and exit of the perpetrator. Note position and whether doors are open, closed, or have been moved.
- Bullet holes, empty shells, and bloodstains. Note and mark location.
- Search for visible and latent fingerprints, plastic and surface footprints, traces of tools, cut telephone wires, traces of teeth, strands of hair, cloth, buttons, cigaret butts, etc.
- Other traces and clues.
- Determine if traces come from victim, murderer, or third party.
- Search terrain about premises or vacant lot, noting vegetation, condition of soil, footprints, etc. Determine movements of victim and murderer.
- Take photographs to show body in original condition with relation to stationary objects, the route of the murderer, etc. The photography should be carefully planned.
- Make diagram of scene.
- Officer in charge should dictate to stenographer complete and detailed description of scene.
- Try to visualize what has taken place, with the aid of the position of the victim, traces of violence, position of bloodstains and weapon, etc.
- Search for clues. Follow to the end. Investigate every theory.
- Preserve evidence.
- Record findings in memorandum book.

In short, the list of things to do provided most of the guidelines that lasted for nearly 50 years—roughly until the dawning of the age of DNA. In some instances, however, it wouldn't be until much later that police would be in a position to benefit very much for many of the precautions they'd been urged to take at crime scenes, because the technology was not yet available to elicit much compelling forensic evidence. Nevertheless, police were already being conditioned and trained to think in forensic terms, and crime scene investigators were already been taught to be careful and thorough.

Tending the injured and the dying after a bomb was thrown during an anarchist-inspired riot in Union Square, New York, March 28, 1908. During such events, securing forensic evidence is always secondary to saving lives. But forensics would, for example, help to convict Timothy McVeigh ninety years later.

Eventually, a field began to emerge to apply the techniques of science to legal

Police search Max Dolinger, the anarchist bomb suspect. Today, ballistics and forensic examination for trace evidence linking him to the bomb would come into play, particularly bomb-making material residues on hands, clothing, or in his place of residence.

matters, in order to furnish the body of evidence necessary to solve crimes and bring criminals to justice. At first, with the exception of fingerprint identification and the use of dental records to identify known missing persons or suspected victims, forensic science was generally of limited use in solving crimes. Prior to World War II, the byword for police detectives remained, "Round up the usual suspects."

But gradually, this began to change. Police still relied on eyewitness reports and confessions to solve most crimes, however the approach to solving crimes eventually became more scientific. Law enforcement officials no longer based all of their actions on oral statements, victims' accounts, and testimonial evidence. In addition, they also looked more for real evidence, physical evidence, such as fingerprints, ballistics, blood, paint chips, and other tangible things. Yet for this evidence to be accepted as valid, it had to be authenticated; it would have to be proved where and how the evidence was obtained, how it was preserved and kept free from possible contamination or tampering.

Criminalistics

Criminalistics is the application of science to answer questions relating to examination and comparison of biological evidence, trace evidence, impression evidence, drugs, and firearms, for legal purposes.

Gradually a new field, criminalistics, began to emerge. In 1973 an influential federal report, entitled *Crime Scene Search and Physical Evidence Handbook*, promoted a more scientific approach for investigating crime, and defined criminalistics as "a profession and scientific discipline which is directed toward the recognition, identification, individualization, and evaluation of physical evidence by application of the natural sciences in law-science matters." The new professional criminalist (crime scene technician, examiner, or investigator) became recognized as a specialist who searched for, collected, and preserved physical evidence in the investigation of crime and suspected criminals.

Today, the criminalist or crime scene investigator typically works for a police department crime laboratory and usually is expected to be available on call 24 hours a day on short notice to

A Montgomery County police officer stands guard over a blood-stained Chrysler, October 3, 2002, Rockwell, Md. Cab driver Prem Kumar Walekar was shot and killed. In the previous 15 hours five people had been shot to death within a ten-mile radius. Forensics quickly identified the same gun as the murder weapon in three of the shootings. The hunt for the Washington Sniper was on.

visit crime scenes, at the front end of a criminal case. A criminalist may type blood, search for drugs, or look for trace evidence at the crime scene. Criminalists are also often called to testify in court about what they have done with criminal evidence. Although the term was coined more than a century ago, it has only been commonly used in the USA since the 1960s, when several four-year colleges and two-year community colleges began to establish concentrations in police science or criminology programs to train personnel to work in some forensic science capacity in law enforcement. Federal grants and loans served to support this activity.

Criminalistics is usually viewed as a sub-branch of forensic science or applied criminology. Since the late 1980s, with police expansion in combating drugs and the advent of new technologies such as DNA analysis, the need for more labs and criminalists has significantly increased. Most criminalists are employed by police departments, but a few work as independent agents or consultants.

As millions of television viewers learned by watching coverage of the O.J. Simpson case, defense lawyers often challenge criminalists at trial by making them testify about matters of contamination, cross-contamination, and chain of custody, in an effort to question the reliability of the evidence found at the crime scene.

Over the years, the care and handling of physical evidence has become much more contested and greater care has had to be taken to ensure that materials have been properly obtained, kept, and managed. Without procedures and documentation to prove that this has occurred, and demonstrated professionalism of the personnel involved, the integrity of the evidence may be successfully challenged. A documented chain of custody must be shown, clearly showing who took the evidence, where they found it, and how they obtained it. Such evidence must be sealed (usually with evidence tape) and marked, to prevent tampering, contamination or substitution. Records must show each and every person who had access to the evidence. The paramount goal is to ensure that the chain of custody is not broken.

Since the 1990s, pressure has increased to ensure that criminalists or forensic scientists are properly trained, certified, and supervised. Criminalists working in crime labs have increasingly been required to undergo proficiency tests in their particular discipline, to ensure that they and their lab are performing up to the standards of the profession. Both forensic scientists and criminalists have also been increasingly expected to be intimately familiar with the governing Standards of Admissibility for Scientific Expertise. They must be scientifically able to record, identify, and interpret minutiae within the parameters of science and law.

Forensic Science

Forensic scientists employ valid and reliable techniques from various disciplines to furnish the body of evidence to support a criminal charge—or the evidence to exonerate an individual or show that no such crime occurred. They are supposed to seek the truth using empirical methods.

Over the last 70 years or so, the scientific fields providing the foundation for forensic work have generally included biology, chemistry, anthropology, and engineering, to name a few. During the last two decades or so, however, a veritable explosion has occurred in the use of forensic sciences in criminal investigation.

Some of this development can be seen in a vast expansion in the number and range of disciplines that have become applied in efforts to solve crimes. Today there are recognized experts and in some instances professional associations, professional publications, and standards in such diverse areas as the following:

Forensic medicine,
Forensic anthropology,
Forensic deception analysis,
Forensic odontology,
Forensic entomology,
Forensic accounting,
Forensic art,
Forensic ecology,
Forensic geology,
Forensic pathology,
Forensic biology,
Forensic psychiatry/psychology,
Forensic nursing,
Forensic dentistry,
Forensic archaeology,
Forensic sculpture,
Forensic limnology,
Forensic botany,
Forensic chemistry,
Forensic imaging,
Forensic neuropsychology,
Forensic mathematics,
Forensic radiology,
Forensic biomechanics,
Forensic seismology,
Forensic linguistics.

Courts and lawyers find themselves confronted with a complex and sometimes bewildering bevy of forensic specialties of greater or lesser stature. Some of them are familiar to legal practitioners; others are not. And the state of the science in each of them keeps changing.

Science vs. Law

Forensic science involves the application of science to legal evidence, yet this can prove to be a tricky fit. The legal process and the scientific method do not necessarily have the same aims.

Science and law sometimes conflict, perhaps because they have different aims. It's said that science exists to find truth, while law exists to serve justice—i.e., to enable people to reach some sort of accommodation. Based on the scientific method and all of its associated rules, scientists can't ignore relevant facts or data; they must consider everything. But lawyers can exclude certain information, in order to resolve a conflict and reach a "just" result; those who administer the law may "go by the book" or use their legal discretion.

Courts have long struggled over how to evaluate or control the use of "science" in law. In 1923, the United States Supreme Court in *Frye v. United States*, established a rule for deciding whether to admit novel scientific evidence. In that case, it specifically rejected the validity of polygraph or lie detector tests, saying in part:

> Just when the scientific principle of discovery crosses the line between the experimental and demonstrable stages is difficult to define. Somewhere in the twilight zone the evidential force of the principle must be recognized, and while courts will go a long way in admitting expert testimony deduced from a well-recognized scientific principle or discovery, the thing from which the deduction is made must be sufficiently established to have gained general acceptance in the particular field in which it belongs.

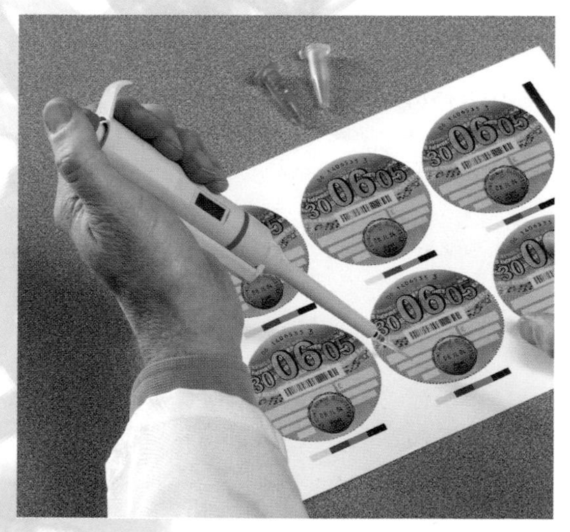

Raman spectroscopy reveals the spectrum caused by scattering of light due to the transition between vibrational and rotational energy levels in molecules. What this actually means is that with a reagent applied over a small area, dyes can be authenticated.

For decades, judges tried to apply this rather hazy "general acceptance" standard and lawyers argued for or against its application in their particular case. However, as scientific advances proliferated and appeared more complex, general acceptance in a particular field seemed more and more difficult to decipher. Furthermore, the evidence standard appeared to discourage the use of valid but innovative techniques that had not yet become generally accepted in the field.

In 1993, in *Daubert v. Merrell Dow Pharmaceuticals*, the Supreme Court adopted a new rule that sought to redefine the standards judges should apply for determining the quality of expert witness testimony. Presently, this opinion still governs the admissibility of scientific evidence in federal court and many state and local jurisdictions that have adopted it.

The plaintiffs in *Daubert* were the parents of children born with birth defects they argued had been caused by the defendant's drug Bendectin. The defense argued that expert witness testimony by scientists who cited studies suggesting it might have happened was inadmissible, because the scientific community as a whole was of the opinion that the drug could not have caused the defects: that there was a "general acceptance."

Following *Frye*, the trial judge agreed and the case was dismissed. The plaintiffs' appeal reached the Supreme Court.

Daubert defined the standards of admissibility more precisely by stating that the Federal Rules of Evidence supersede "general acceptance" tests for admissibility of novel scientific evidence. Under the present standard, a trial judge must screen scientific evidence to ensure it is relevant and reliable, and to do this the focus must be solely on principles and methodology, not on the conclusions they generate. *Daubert* gives broad discretion to trial judges and instructs them to consider at least four factors when determining admissibility. These include:

- whether the theory or technique can be tested;
- whether the science has been offered for peer review;
- whether the rate of error is acceptable; and
- whether the method at issue enjoys widespread acceptance.

The ruling changed the admissibility of expert testimony in two basic ways. It based the test for admissibility of evidence on "scientific knowledge"—not merely on whether it was generally accepted in a particular field, but whether proof of reliability of a technique or scientific method could be established. And it gave trial judges the power to determine this reliability.

Under the new system, courts are required to determine whether a particular expert witness is qualified to testify. If he or she is qualified, they can be allowed to provide specialized information that may assist the court in interpreting the factual evidence.

Experts are not fact witnesses like eyewitnesses to a murder or witnesses who observed the defendant acting strangely at a particular time and place. But a forensic pathologist is both a fact witness and an expert witness—he or she testifies both to the facts of the autopsy findings and interprets their meaning.

At the Crime Scene

Under the new paradigm, crime scenes assumed paramount importance because it was recognized that that was where the evidence of the crime could usually be found, provided it was properly handled.

The crime scene was where the fatal shooting occurred, as evidenced by the body lying in a pool of blood, the ejected shell casings, the bullet holes, and the shattered glass. It was where the burglar had pried open a window and left his fingerprints on a light switch. Bloodstains left at the site, or suspicious hairs found on the victim's body could help to determine how the crime was committed and who committed it. Tire tracks or carpet fibers present at the site could prove useful in the investigation. Such physical evidence might assist the police to establish the criminal's *modus operandi*, link the crime or the criminal to other persons or places that might prove revealing, or help to prove or disprove critical facts of a witness's account. Such clues might provide investigative leads. Indeed, the crime scene might yield evidence that ultimately would lead to identification and conviction.

The trick was—and is—how to scientifically recognize, document, capture and preserve, identify, compare, individualize, and evaluate such evidence, so that it would help to solve the crime and hold up in court. Without a properly trained investigator to mine effectively the clues present at the crime scene, the crime laboratory would not obtain the stuff on which to conduct its analysis, the medical examiner would not be able to reach his or her final conclusions effectively, the prosecutor would not be convinced to bring formal charges against the suspect, and the jury would not convict.

Photography

Crime scene photography has of course become more sophisticated, precise, and instantaneous over the 170 years since the daguerreotype. It takes quite a leap of the imagination to consider how crimes were investigated and criminals pursued and prosecuted *without* the camera.

Soon after its invention in the mid-nineteenth century, photography was harnessed to assist in gathering evidence to help solve crimes. It became one of the first and most

Rogues Gallery, NYPD, 1909. The Metropolitan Police, London, had collected photographs of prisoners from prison governors since 1862, and employed the first police photographer in 1901.

Photographer with what looks too big a camera for a Box Brownie, (and it's a little too early) around 1910; the first mugshot was probably taken in Belgium in 1844. The first photograph used as evidence in a court of law was taken by Louis Daguerre himself in 1839, in a divorce case. Husband and floosie caught outside the Paris Opera House: divorce granted.

indispensable forensic tools. Police developed uniform mugshots and took pictures of crime scenes and evidence to help memorialize their investigations. Sometimes their precise and graphic portraits of the murder victim were used as exhibits that helped to sway the jury to convict or condemn the accused killer. But most early crime scene photography had limited forensic value.

By the early part of the twentieth century, news photographers also often worked alongside the police to depict the corpse, the blood, and other shocking details of the crime scene, sometimes to the detriment of the investigation, as the journalists paraded all over the evidence and dropped cigarette butts on top of blood spatters, or accidentally kicked away spent bullets and other clues.

By the 1950s the typical image of a murder scene featured several well-dressed and middle-aged police detectives wearing fedoras and business suits, who knelt around a victim's body exchanging observations and occasionally jotted down notes and sketches. In the background, pairs of other detectives often stood off to the side, talking among themselves. A few uniformed cops kept back the crowd.

Then a plainclothes cop would come and arrange his camera on a tripod situated about three feet or so from the body. The police photographer would step onto a small metal box and bend down over the camera, covering his head with a black silk sheet. A nearby light illuminated the body and other items within a radius of a few feet. Later the photographer would systematically shoot the scene from different angles, stopping to jot down a note about what he had recorded.

Over time the crime scene photographer became more alert to other potential evidence, such as a half-eaten apple or a broken bottle lying near the corpse. The police inserted measuring sticks or other items into the picture to show scale. They also took care to record accurately any scratches or other wounds on victims or subjects. Photographers were instructed to document any footprints leading from the shooting and to record minutely each and every blood spot.

Photography also became an essential forensic tool in the morgue and the police laboratory. Photographic documentation creates a permanent, detailed, and accurate historical record of the crime scene or autopsy and supplements witness accounts and other data to help the police to reconstruct what happened.

The practice of forensic photography has changed dramatically over the decades. Today's police photography requires the police to first remove all nonessential personnel from the scene. Photographers are trained to obtain an overall (wide-angle) view of the scene to locate spatially the specific scene in relation to the surrounding area, and also to photograph particular areas of the scene to provide much more detailed views of spaces and objects. This is done from several different angles to provide various perspectives that may uncover additional evidence.

Dead bodies must be photographed full-face for identification purposes and the body and clothing must be shown in detail.

The photographers use tripods to keep the camera steady and exactly perpendicular to the subject, as well as measuring sticks to show size. They utilize forensic light sources that direct brilliant, narrow beams on the subject, and their cameras are equipped with color filters that can utilize ultra-violet to highlight fingerprints and stains. Their lenses and film produce extraordinarily sharp images. Despite the rapid proliferation of digital cameras, which offer greater speed and flexibility, crime scene photographers tend to use 35mm analog cameras because they provide greater accuracy and proof that an image has not been altered.

Video photography has also become a staple of police investigation, in part because it can record so much more information about the police response. As will become clear later in this book, such evidence is not always to the advantage of the prosecution; police mistakes can be recorded just as accurately as correct procedure.

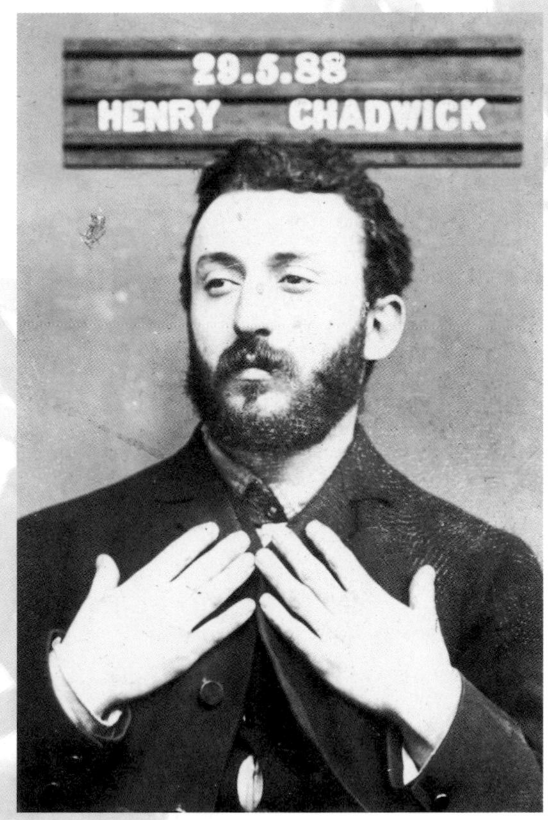

Police portrait taken in Manchester, England, in 1888. The inclusion of the hands in this way perhaps as a further means of identification is interesting. A written record of convictions would accompany the portrait.

From Lindbergh to O.J.

The evolution of crime scene investigation and forensic science can be illuminated by looking at two of the twentieth century's most sensational cases. From 1935 to 1995—a course of 60 years—American crime investigation naturally underwent many dramatic changes: the greatest of these was professionalism. In response to the problem of gangsterism during the Great Depression, President Franklin D. Roosevelt helped to create the modern Federal Bureau of Investigation as a model police force. He also scheduled a National Crime Conference in 1934 that was designed to promote the issue of police professionalism for small police departments as well. Emphasis was placed on training personnel to help them effectively cope with criminal operations. FBI Director J. Edgar Hoover envisioned one of this agency's roles as being to "provide assistance in training through methods of scientific criminal detection and other law enforcement activities."

Gangster Louis "Lepke" Buchalter, center, photographed in 1939 four years from the electric chair at Sing Sing, is handcuffed to J. Edgar Hoover on the left. Hoover was always willing to claim credit for himself and the Bureau, even when none was due, as in the Lindbergh kidnap case.

Police began to utilize two-way radios, dispatchers, automobiles, fingerprinting and ballistics X-ray equipment. More states continued to form and bolster their own state police departments, complete with their scientific crime labs and training academies.

The evolution of forensic science from this early start in the 1930s to a period of revolutionary change brought on by the advent of DNA in the 1990s is bookended by two landmark cases: the Lindbergh kidnapping of 1932–35 and the O.J. Simpson case of 1994–95—both of which were called the "crime of the century." There were different outcomes, not only in the cases themselves, but also for the science of forensics.

Or perhaps more accurately in the later case, that uneasy fit mentioned earlier between science and the law, between truth and justice, was strained; some would say, to breaking point. Others would point to the triumph of the justice system.

Microscopists: Lindbergh Under the Microscope

Mr. and Mrs. Charles Lindbergh leaving church in Washington D.C. some time in the late 1920s. When Lucky Lindy flew the Atlantic in 1927, he became one of the most famous men on Earth. Which can be a dangerous thing. Handwriting analysis would link Bruno Hauptmann to the ransom notes, and crucially—anyone can write a ransom note—pine wood put the carpenter at the scene.

The Lindbergh baby's kidnapping ushers in new forensic attention to handwriting analysis and microscopic mark comparisons.

One of the greatest forensic, legal, and media cases of the twentieth century, the Lindbergh kidnapping, continues to generate intense controversy, mainly owing to issues surrounding the fairness of the trial. For sheer notoriety, no other high-profile forensic case came close until the JFK assassination and O.J. Simpson. But most experts agree about the guilt of the lone defendant who was convicted and executed.

The young and dashing aviator, Charles A. Lindbergh, was one of America's favorite heroes based on his first-ever solo flight across the Atlantic. So when his only child was kidnapped for ransom from Lindbergh's Hopewell, New Jersey, home on March 1, 1932, the world's press reported every new development.

Police and reporters swarmed over the crime scene, looking for clues or any dramatic or sensational detail that would generate a scoop. They trampled all over the grounds and through the house, leaving fingerprints and footprints everywhere. Newspaper photographers snapped pictures of every conceivable piece of evidence and participant they could find.

One of the hottest early reported clues, found by Colonel Lindbergh himself, appeared in the form of a crude ransom note that analysts quickly attributed to a poorly educated person of European, probably German, extraction. Another key piece of evidence was the homemade ladder the culprit or culprits had left behind. There was also an old chisel and a few muddy footprints.

After more communications and the payment of a $50,000 ransom, hopes were dashed with the discovery of the baby's decomposing body in nearby woods. The cause of death was officially classified as a fractured skull or blunt force trauma—possibly related to a fall from the ladder—yet there was no autopsy by a forensic pathologist or medical examiner. (A latter-day study of the medical records by Dr. Michael Baden, one of the world's leading forensic pathologists, however, concluded that the baby probably died from suffocation, when the kidnapper tried to keep the infant from crying out during the crime.)

THE SCIENTISTS

James E. Starrs—historical forensic sleuth, "the bone hunter," Professor of Law & Forensic Sciences at George Washington University Law School, expert in scientific evidence who has investigated several historic cases, including the Lindbergh kidnapping, Sacco & Vanzetti robbery-murder, the Boston Strangler, the identification of Jesse James, and assassination of Huey Long.

Police sought help from a wood expert, Professor Arthur Koehler of the Department of Agriculture's Forest Products Laboratory, to help them examine the homemade ladder. He determined that the ladder's top section (rail 16) was made of southern yellow pine with knots and distinctive raised grain patterns. It had also been hand-planed along both edges and mysteriously contained 4 extraneous holes on its face. That particular wood was often used for flooring. Microscopic inspection revealed distinctive plane marks that were traced to a sawmill in South Carolina. Eventually the investigators located a storage bin in the Bronx that contained lumber from the same mill lot and laboratory analysis revealed it matched the wood used in the ladder. But records did not disclose who purchased the material.

Then an informant, gas station manager Walter Lyle, reported he had jotted down the license number of a suspicious man who had passed him a $10 gold certificate containing one of the serial numbers included in the ransom. The police traced the car to Bruno Richard Hauptmann, a German immigrant of modest means who also had a criminal record. He lived in an upstairs apartment in a two-story house in the Bronx.

In Hauptmann's attic they discovered a missing floorboard and the wood in the floor matched the fir used in the ladder; it even had four nail holes that lined up with where the nails had been sunk in Hauptmann's attic joist. The ladder wood also contained several distinctive marks that appeared on the attic floorboards. Police also discovered $14,000 of the ransom money hidden in a tin can in Hauptmann's garage.

STATE OF NEW YORK—DEPARTMENT OF TAXATION AND FINANCE—BUREAU OF MOTOR VEHICLES
APPLICATION FOR REGISTRATION
1934 PASSENGER VEHICLE
NOT USED FOR HIRE
Use Special Blank on and after July 1st
1. Print Name of owner RICHARD HAUPTMAN IV
HAUPTMANN'S OWN LETTERED SIGNATURE ON AUTO REGISTRATION CARD
RICHARD HAUPTMAN IV
RICHARD HAUPTMANN
SAME SIGNATURE RECONSTRUCTED FROM LETTERS CUT OUT OF KIDNAP NOTE

MR. CHAS. LINBERG.
YOUR BABY IS SAFE BUT HE IS NOT
USING NO MEDICINES. HE IS EATING
PORK CHOP. PORK AND BEANS JUST WHAT
WE EAT. JUST FOLOW OUR DIRECTION
AND HAVE ONE HUNDRED THOUSEND
BUCKS READY IN VERY SHORT TIME
THATS JUST WHAT WE NEED
YOURS B. H.

Top, Hauptmann's "signature" on an auto registration card, a ransom demand (below) signed "B. [Bruno] H." Criminal mastermind he wasn't.

Besides the microscopic analysis of marks in the wood, the Lindbergh case featured the first major attempt to use handwriting analysis in a murder case. Six specialists in the newly emerging science of questioned documents all testified that Hauptmann had written the ransom notes.

Although some writers at the time questioned some of the wood and handwriting evidence as faulty science, and many more observers wondered if Hauptmann had been assisted by any unidentified accomplices, on February 13, 1935, Bruno Hauptmann was convicted of first-degree murder. He went to his death in the electric chair still protesting his innocence and never incriminating anyone else.

The handwriting analyisis was surely impressive: but it only linked Hauptmann to the ransom notes and not to the kidnapping. The crucial evidence is the ladder. If, as some have suggested, this evidence was fabricated, then the case collapses. But this seems a conspiracy theory too far.

Burglar, armed robber, and illegal alien Bruno Richard Hauptmann. Hauptmann was not caught by forensic analysis but by a suspicious gas station manager who wrote down Hauptmann's license plate.

Crime Scene Investigation: O.J. Simpson

The O.J. Simpson case assembled a "dream team" of top lawyers and forensic experts who put the LAPD's shoddy crime scene investigation under the microscope to show that much of the crucial scientific evidence had been contaminated or didn't add up.

When Nicole Brown Simpson and Ronald Goldman were found slashed to death at the entrance to Simpson's home at 875 South Bundy Drive in Los Angeles on the night of June 12, 1994, suspicion quickly fell upon her estranged husband, O.J. Simpson, the famous actor and former football star. He had previously been involved in domestic violence incidents with her and some of his ex-wife's relatives immediately told the police that he had "finally killed her." Crime scene investigators started swarming over the site and LAPD detectives also began looking for O.J. Simpson at his upscale compound five minutes away on Rockingham Avenue.

By the time an LAPD detective contacted O.J. Simpson by telephone to inform him that his wife had been "killed," he was in Chicago on a business engagement and he rushed home. Police

Lead defense attorney Robert Shapiro talks with defendant O.J. Simpson during a pretrial hearing on evidence seized from Simpson's Ford Bronco vehicle, October 6, 1994.

already had found several incriminating clues, including blood on the door of his parked Ford Bronco as well as blood drops leading into his mansion. Detective Mark Fuhrman reported finding a bloody glove at the crime scene as well as a matching glove on the south service pathway to O.J.'s home on Rockingham. (Testing would later show blood that was consistent with O.J., Nicole, and Goldman. The glove also contained African-American limb hairs and hairs consistent with Goldman and Nicole, as well as blue-black cotton fiber consistent with the clothes that O.J.'s roommate Kato Kaelin said he saw Simpson wearing on the night of the murders.)

Shortly after O.J.'s return from Chicago, the police noticed his bandaged hand and began questioning him, photographing the wound, taking samples of his blood and hair, and finally they arrested him for the murder.

LA detective Mark Fuhrman found both gloves, many of the blood drops in the dark, and trespassed on Simpson's estate without a warrant.

But Simpson was unlike other defendants. He was rich. And he began using some of his resources to hire the best legal defense team ever assembled in the United States—a group with so many legal stars that the news media dubbed it the "Dream Team."

The new lead counsel, Robert Shapiro, and his colleagues, Johnnie Cochran, F. Lee Bailey, Alan Dershowitz, and several others, quickly realized the role that forensic evidence would play in the case, and they in turn brought in several of the nation's premier forensics experts to assist in the defense. They included: Barry Scheck and Peter Neufeld, two New York-based lawyers specializing in DNA; Dr. Michael Baden of New York, regarded as the top pathologist and medical examiner; renowned criminalist Dr. Henry C. Lee of Connecticut; Herb McDonald, the world's leading blood pattern expert; Chuck Morton, a famous trace evidence expert; crime scene expert Larry Ragel, and several others. Most of the experts retained by the defense were among the best-known authorities in their fields, and were usually employed by the prosecution; their integrity and credibility was well established.

As it would turn out, unlike virtually any other high-profile case tried in Los Angeles, this high-powered defense team would have at its disposal almost as many resources as the prosecution, amounting to enough talent to make the Simpson case one of the greatest American courtroom battles. Ultimately the trial would involve 126 witnesses and 857 pieces of evidence. It would receive the most intensive coverage of any criminal trial in history, with live televised broadcasts

A remarkable moment of courtroom theater: O.J. Simpson demonstrates that the Aris gloves apparently do not fit. One more strong seed of doubt was planted in the minds of the jury. All the forensic evidence was steadily undermined.

of the proceedings and endless commentary by scores of talking heads. Networks and supermarket tabloids paid huge fees in exchange for eyewitness "scoops."

Before Simpson, crime scene investigators for the LAPD (and many other major police departments) were accustomed to handling much of their blood and other biological evidence much more casually, even sloppily. Cops had not been sufficiently trained to deal with DNA.

But as the Simpson trial made clear, the power of scientific evidence can cut both ways: on the one hand, DNA can establish guilt or innocence more clearly than anything else; on the other, mistakes by the police at any one link in the evidentiary chain, either by failing to properly gather or store blood swatches and other evidence, or by bungling its handling in the laboratory, can destroy even an open-and-shut case. If nothing else, the jury's "not-guilty" verdict should have shaken big-city police departments, especially LA's, out of their small-time forensics complacency.

One of the trial's most dramatic moments involved the bloody gloves that Detective Fuhrman said he had found. With help from the FBI, the prosecution established that Nicole had purchased two identical pair of Aris leather gloves, size extra large, and records showed the gloves were very rare. The prosecution claimed the gloves had belonged to O.J. and that he had worn them to commit the murders.

But Christopher Darden, the assistant prosecutor, allowed Simpson to demonstrate whether the gloves actually fit. Experts had already claimed that the blood and other material on them would not have caused them to shrink. But when Simpson attempted to try them on in full view of the jury, they appeared to be too small for his large hands. And as defense lawyer Johnnie Cochran later concluded in his summation: "If the gloves don't fit, you must acquit."

In theory, the prosecution should have been able to ensure that comparisons of DNA from blood found at the crime scene, Simpson's car, his house gate, and a sock found in his home, all proved that he had committed the murders. But in light of questions raised about the police department's sloppy handling of the evidence, and the racial attitudes of some of the officers, jurors were left questioning how the DNA might have ended up there. In the end, such questions created doubts that resulted in Simpson's acquittal.

Looking back on what transpired in the O.J. Simpson case, experts identified some of the LAPD's most glaring forensics mistakes as follows:

- The defense repeatedly used some of the LAPD's own crime scene photography, both still pictures and videography, to reveal mistakes in police handling of evidence, as well as to show that some of the alleged evidence had never been photographed or did not appear at the original crime scene, thus raising questions about whether it had been planted.
- Crime scene investigators had failed to collect pieces of crucial evidence. For example, police photos of Nicole Simpson's corpse revealed blood spatters on her upturned back that analysts later concluded must have originated from someone else—perhaps her killer. But police investigators failed to take a genetic sample of the blood before turning the victim over, thereby contaminating any possible samples.
- Renowned criminalist Dr. Henry Lee testified for the defense that he found a new trail of seven blood drops leading away from the killing scene that hadn't been visible in the LAPD's poor-quality pictures given to him for review. He also criticized the quality of LAPD laboratory microscopes and other equipment.
- The prosecution presented compelling photographic evidence claiming to show O.J. Simpson wearing the same type of extremely rare, size 12 Bruno Magli shoes, with soles that matched bloody footprints found leading from the bodies. But this evidence was challenged as a fake. (The case had already been marked by a highly publicized doctored image in the form of an altered image of O.J. Simpson that had been published on a news magazine cover.)
- Police took more than two weeks to remove blood from a fence, a lapse that left many observers wondering if it had been planted. Scientists pointed out that the long outdoors delay had fatally damaged the evidence.
- Among the other items of useful evidence the police failed to preserve or record was a dish of melting Ben & Jerry's ice cream found in Nicole's home—evidence that could have been extremely helpful in pinpointing the time of death.

- The collection of evidence at the crime scene was incredibly sloppy. John Gerdes, M.D., a DNA expert and the clinical director of Immunological Associates of Denver (IAD) who testified for the defense, watched the police video of the crime scene with the jury and pointed out many problems with evidence collection, including LAPD Criminalist Andrea Mazzola's repeated failure to change her gloves after collecting evidence. The video showed Mazzola swabbing up blood drops while leaning a gloved hand on dirty ground, touching tweezers with the same hand, then using the same tweezers to manipulate a bloody swatch. She also placed wet swatches in plastic bags where Gerdes said bacteria could grow and "cleaned" the tweezers by merely wiping them with clear water—a procedure that was not likely to remove the DNA.
- LAPD Criminalist Mazzola testified she took swatches from bloodstains at Simpson's Rockingham estate, placed them in paper envelopes, and put them in the crime lab truck. Her testimony confirmed that she had not put them in proper containers or immediately put them under refrigeration to prevent their degradation and contamination.
- Other personnel working at the crime scene were shown to have operated without wearing the required gloves, hairnets, booties, and other protective equipment.
- LAPD criminalists were shown to have collected hair, fiber, and other trace evidence in a sloppy way by putting all such evidence into the same container, thereby rendering it contaminated.

Blood on the path at Brentwood, photographed June 15, 1994. For three days, prosecution witness Dr. Robin Cotton explained how DNA tests work; and how the odds that the blood found near the bodies was not Simpson's were one in 170 million.

- To facilitate access to the blood-covered crime scene, police used bath towels from the house to mop up large quantities of blood lying in the entranceway. Some workers had actually dumped some of the towels, used gloves, and other debris on top of a victim's body. They had also stepped all over the bloody surface and tracked blood from place to place.
- One of the worst mistakes was that the victims' bodies were left lying in the open air for hours, without being examined by a medical examiner. He was not even notified until ten hours after the bodies were found.
- Still-wet blood was belatedly found on socks in O.J.'s home (but not photographed close-up there) and when tested the blood appeared to consist of a mixture of O.J.,'s, Nicole's, and Goldman's. But the defense

was able to challenge this evidence by questioning how it could have remained wet for such a long time, and the defense's forensic toxicologist testified he found the preservative EDTA (ethylenediaminetetraacetic acid), which prevents blood from coagulating, in the bloody socks and in a stain on the back gate of Simpson's ex-wife's condominium. The toxicologist concluded there were only two possible sources for the EDTA—from a blood sample tube or through contamination of the bloodstains in the laboratory. Either way, doubts were raised about the integrity of this blood evidence. (Dr. Robin Cotton, laboratory director of Cell Mark Diagnostics, Germantown, Md, had testified that tests showed that blood on the socks had the same genetic fingerprint as Nicole's, characteristics that matched only one in 9.7 billion Caucasians; it was her blood. But her testimony was gradually drained of its authority through day after day of highly technical questioning, not only of the integrity of the evidence but also the methods and principles of DNA testing itself.)

- Detective Phillip Vannatter left the police station with a vial of O.J.'s blood in his back pocket, then he drove to the Bundy house and walked around the crime scene still holding the blood sample, until he finally handed it off to a criminalist for testing. This was a serious mistake because he did not handle the sample properly from a scientific perspective and he also may have compromised its legal status.
- After the murders, the football star claimed that cuts on his hands had been caused by a broken glass in his Chicago hotel room, but police failed to preserve and test the glass.
- Police failed to record and account for the precise amount of blood taken from O.J. and later used for various testing, thereby giving rise to suspicions that some of it may have been used to plant evidence against him.
- The coroner failed to analyze and record properly the contents of the victims' stomachs and thereby compromised his ability to estimate the time of death. In all, Dr. Baden noted at least 16 mistakes in the autopsy.

In the end, the O.J. Simpson case educated not only the police, but also the world about some of the power, complexities, and pitfalls of forensic science. Everybody learned that forensics can cut both ways.

ON THE STAND: PATROLMAN RISKE

Excerpts from the trial testimony of Patrolman Robert Riske, the first LA police officer to arrive at the Simpson/Goldman murder scene:

MR RISKE: As we arrived at the scene we were flagged down by two witnesses and a dog. They directed us to 875 and they said there was a dead lady on the walkway.

MS. CLARK: And after they told you that what did you do?

MR RISKE: My partner and I crossed the street and went to the walkway and what I observed was a female white in a black dress laying in a puddle of blood on the walkway.

MS. CLARK: When you approached this area, can you describe the lighting for us, sir?

MR RISKE: There is—there is really no street lighting. There is overhanging trees over the walkway. There was lights on in the residence, but there is a fence that goes in front of the windows, so the lighting was poor.

MS. CLARK: And was there a lot of foliage, a lot of bushes and trees in that area?

MR RISKE: Yes.

MS. CLARK: And what did you do when you stood there with your flashlight?

MR RISKE: Just turned the flashlight on and saw the body and went back and walked to the—requested a supervisor, additional units and an ambulance and then we went back and talked to the witnesses.

MS. CLARK: All right. What did you do next?

MR RISKE: We went back to the scene, we approached the body of the female, and as we got probably two feet from her body, we dicovered the body of a male white lying against the north fence.

MS. CLARK: Okay. How did you—how did you get up to the woman's body? What did you do?

MR RISKE: My partner approached on the grass and I approached walking through the plants right there, staying to –

MS. CLARK: These bushes here on the left? [indicating]

MR RISKE: Right. Staying left of the walkway.

THE SCIENTISTS

Barry C. Scheck—New York criminal defense attorney specializing in DNA issues, co-founder with Peter Neufeld of the Innocence Project, Yeshiva University. He burst into the national limelight through the O.J. Simpson case and has helped to marshal DNA evidence to obtain exonerations for wrongfully convicted persons throughout the U.S. Co-author with Neufeld and Jim Dwyer of the best-selling book, *Actual Innocence*.

MS. CLARK: And your partner, where was he?
MR RISKE: He was on the grass to the left of the foliage. You can't really see in this picture.
MS. CLARK: So you were in the bushes and he was to the left of you as we face the photograph?
MR RISKE: Right
MS. CLARK: Did you step on the walkway?
MR RISKE: No.

ON THE STAND: DR. LEE

Dr. Henry C. Lee questioned about blood at the Simpson-Goldman murder scene:

MR. SCHECK: Well, let me ask you this, Dr. Lee. In a closed environment, closed-in environment with hand-to-hand combat, with multiple stab wounds, with blood stains in different places indicating multiple contact smears with vertical droplets in the areas of the different multiple contact smears, with other blood spatter cast off in different directions, with the key in one area, beeper in another area, in that kind of struggle, do you have an opinion as to whether or not an assailant or assailants would be covered with blood from the struggle?
MR. GOLDBERG: Misstates the testimony, calls for speculation.
THE COURT: Overruled.
MR. GOLDBERG: Incomplete hypothetical.
THE COURT: Overruled.
DR. LEE: Yes.
MR. SCHECK: What is that opinion?
DR. LEE: In theory, should have some blood.

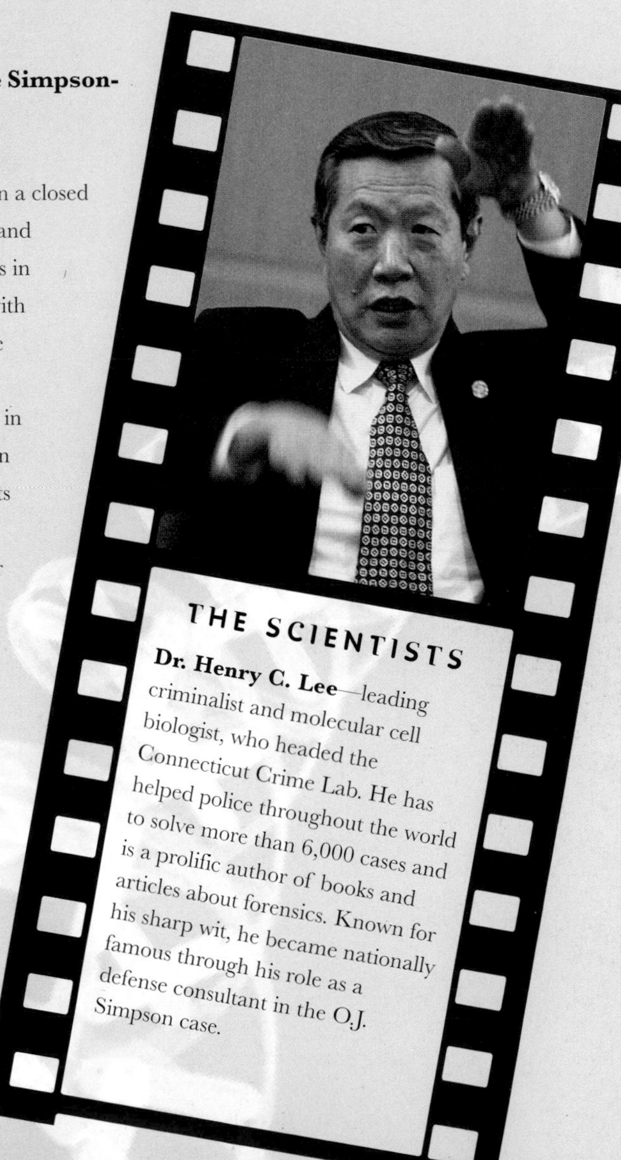

THE SCIENTISTS

Dr. Henry C. Lee—leading criminalist and molecular cell biologist, who headed the Connecticut Crime Lab. He has helped police throughout the world to solve more than 6,000 cases and is a prolific author of books and articles about forensics. Known for his sharp wit, he became nationally famous through his role as a defense consultant in the O.J. Simpson case.

Crime Scene Search

In a post-O.J. world, crime scene investigators are schooled to be as careful and methodical as possible.

From the 1930s to the 50s, a street murder in a big city was apt to attract a horde of ghoulish newspaper photographers—some of whom waded through puddles of blood to snap intimate shots of bullet-riddled bodies—crowds of curious bystanders, and cigarette-smoking cops milling around the evidence like delegates on the floor of a political convention. The most famous of all the snappers, Weegee (Arthur Felig, 1899–1968) of the *Daily News,* prided himself on getting closest to the action, and doing it faster than anyone else. Back in the old days, some news photographers bent over and grabbed a piece of evidence as if it were a flower there for the picking, because it would look good on the front page. Bystanders swarmed around Dillinger's corpse in Chicago, sopping up some of his blood with newspapers to take away as relics.

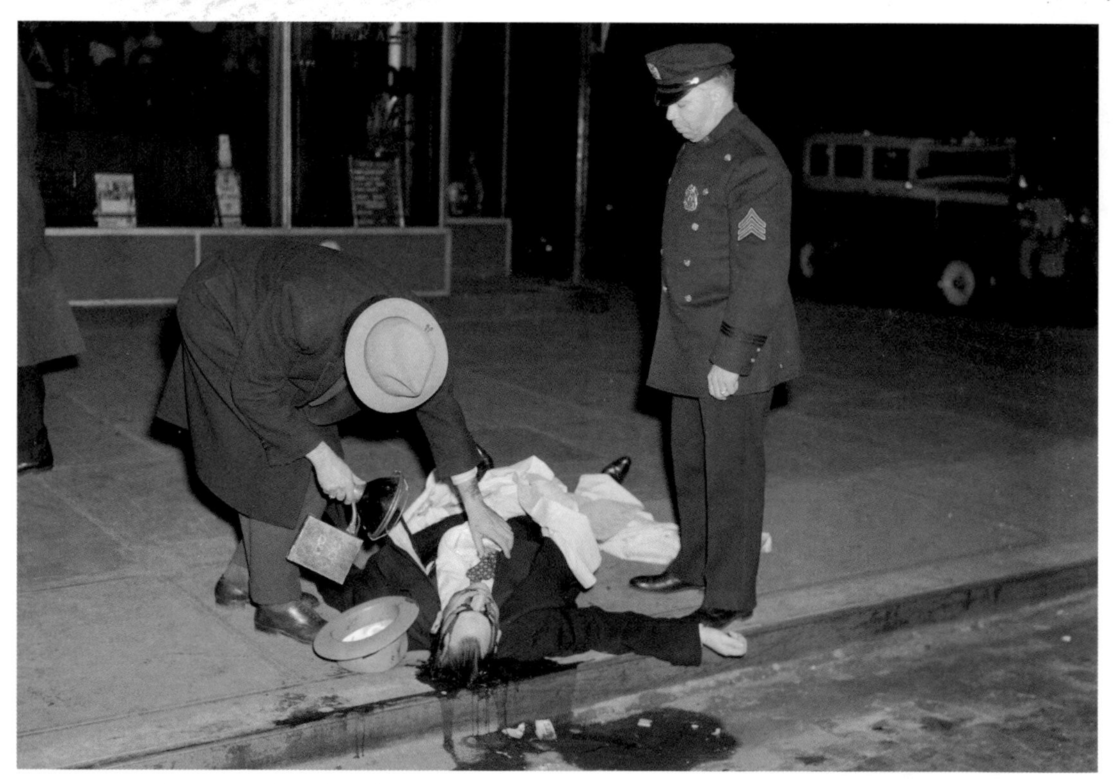

Legendary news photographer Arthur Felig (Weegee) messes with the crime scene, December 9, 1939. David "the Beetle" Beadle lies in front of the Spot Beer Tavern in Manhattan. Weegee would stamp the back of his prints, "Credit Photo by Weegee the Famous."

Today's major crime scenes are sealed off like tombs or quarantined places, off-limits to all but a chosen few. Their investigation has become a major production, complete with specially equipped vans and technicians in space suits. Years ago, the cop assigned to crime scene investigations was the "guy who'd fallen off his motorcycle," somebody who needed to perform a lighter duty. But today the assignment requires a much higher level of education, training, motivation, and support. Today's CSI personnel are much more thorough and careful than their predecessors ever were. They have to be.

In 1998 a policeman in Poughkeepsie, New York, discovered a woman's rotting corpse inside the home of Kendall Francois, and investigators immediately sealed off the area. The stench was indescribable. Further searching turned up another body and the police kept the site closed off to all but the forensic team. In all they found the remains of eight murdered women. The investigators kept the scene sealed for 49 days, or more than eight weeks. Their work prompted a guilty plea.

According to one experienced homicide investigator, "There's just a lot of work to be done. You can never be too careful. I like to say, 'Folks, it's time to roll up your sleeves and loosen your tie, get ready for the long haul. Because we're going to be here for a while.' "

ROLE OF FIRST RESPONDERS

The scandal of the O.J. Simpson debacle taught everyone that one of the most important aspects of securing the crime scene is to preserve the scene with minimal contamination and disturbance of the physical evidence. DNA has changed everything. As a result, police from the first responding officer to the members of the crime scene investigation team are trained to be extremely cautious and careful. One misstep can foul up a case.

Once an initial responder has taken appropriate safety procedures to control any dangerous situations or persons, the next responsibility is to ensure that emergency medical attention is provided where needed while minimizing contamination of the scene. Then the police have to quickly secure and control persons at the scene, removing anyone who doesn't belong there and keeping any suspects and witnesses secure and separate from each other. This can prove more difficult than it may seem, but it is extremely important, particularly because if it is not done it can lead to more contamination of the evidence. Potential witnesses may fashion their stories into a common, mistaken account. Bystanders who are not witnesses or suspects need to be removed from the scene. Because some who appear to be witnesses may later prove to be suspects, the police need to exercise special care to keep them away from the physical evidence as well.

Defining and controlling the boundaries of a crime scene in order to protect the evidence requires common sense and experience. Some boundaries may extend only to the street outside the house where the crime was discovered; others may stretch for acres around the bombsite. The initial responders need to try to establish basic boundaries and potential points of entrance and exit. They also need to erect barriers—patrol cars with flashing lights, armed guards, cones, ropes, signs, and barrier tape—to keep away intruders. Documentation needs to be started to record the identity of every person at the scene and the time that anyone leaves or arrives. Everyone within the perimeter is prohibited from smoking, eating, going to the bathroom, or moving or unnecessarily touching anything.

Upon the arrival of the investigator in charge, that person must formally assume control of the crime scene. This requires first receiving a briefing from the initial responding officer and other personnel. Any documentation is then turned over to the new officer in charge.

This documentation should include a report detailing the conditions upon arrival—whether the lights were on, the window shades were drawn, any odors or sounds that were detected, the temperature and weather conditions, items that were noticed such as guns or burglar's tools.

WALK-THROUGH

After meeting with the first responder and evaluating the safety, search and logistical issues posed at the site, the investigator needs to make an initial scene assessment. Communication may need to be established between multiple sites. The police also have to set up a secure area for temporary evidence storage and possibly a communications center.

Another important duty is to determine what investigative resources will be needed. Since the 1990s, crime scene investigation has gotten much more labor-intensive and the cost of training, maintaining, and equipping its troops has become substantial. Many of the supply items can only be used once. Not every serious crime will require a major crime scene investigation response.

Victims of a fire in Monroe Street, New York City, 1913. Arson, or another tragic accident like the Triangle Shirtwaist factory fire two years before that killed 146? Forensics procedures were not yet up to the task of finding out.

The most intensive crime scene investigation entails extensive equipment and technology, including sophisticated lighting devices and enhancement aids that can enable the searcher to detect many more things that he or she couldn't have seen in years gone by.

Despite this, solving many crimes has become more difficult than it was in previous eras. The clearance rate for homicide, for example, was only 62.6 percent in 2004—more than 25 percent worse than what it was 40 years ago. The FBI's Uniform Crime Reports for 2004 further revealed that the percentage of crimes cleared by arrest was even worse for most other crimes. For aggravated assault, it was 55.6 percent; forcible rape was only 41.8 percent; robbery was 26.2 percent; larceny-theft was 18.3 percent; motor vehicle theft was 13.0 percent; and burglary had a clearance rate of only 12.9 percent. This hardly offered a glowing testimonial for crackerjack police work.

After making his or her initial scene assessment, the investigator conducts a walk-through to provide an overview of the entire crime scene, to identify any obstacles to evidence integrity. The investigator takes extensive notes and photographs to make a permanent record. In a major crime case, the investigator must spend considerable time doing this documentation before he or she even begins to collect and assess the evidence.

Based on this assessment, a specialized crime scene investigation team may be brought in. Said one forensics expert in an interview for this book:

> "Going in to search the scene of a major crime today is like going into the operating room. The technicians have to dress up like surgeons and take all sorts of precautions to prevent contamination."

Contamination control and preventing cross-contamination is essential to maintaining the safety of the personnel and the integrity of the evidence. Fear about crime scene contamination is a major reason why even police brass and DAs are nowadays kept away from major crime scenes, whereas in the old days they liked to make an appearance. In order to enter a major crime scene today, some departments require everybody to provide a DNA sample. DNA has also become a growing labor issue in some departments, as some unions have resisted requirements of their members to submit to DNA testing.

Since the O.J. case, crime scene investigators have been schooled to follow established entry and exit routes, designate a secure area for trash and equipment storage, use personal protective equipment such as rubber gloves, masks and secure suits, and utilize single-use equipment for all biological samples. Their documentation requirements must also be rigorous and closely supervised to ensure evidence integrity. Investigators need to prioritize their evidence collection to prevent loss, destruction, or contamination, particularly because some types of evidence are highly perishable and extremely fragile.

Only after all of these steps have been taken do the investigators actually get down to collecting the evidence—taking swabs, lifting fingerprints, removing carpet fibers, obtaining reference samples and the like. Each collected item must be carefully extracted, marked for identification, documented, and placed in the chain of custody. Electronic recording evidence must be properly secured. Each piece of evidence must be placed in an appropriate container or package to ensure that it is properly protected, preserved, and stored for appropriate action by the laboratory or evidence bureau.

Extreme care must be taken with biological samples. DNA analyses have become better and faster over the years. Pretty soon police will be able to get the results instantaneously from the crime scene, without even relying on the lab. Nowadays, the need to know has become so instantaneous that crime scene video may be transmitted directly from the crime scene back to headquarters or databases. Investigators exchange images and other data via email sent directly from their laptops. At the crash site of TWA Flight 800, the police used digital radiology to search for 100 dental records in 20 minutes to identify the victims.

The downside of the digital evidence revolution is that some defense attorneys may allege that the evidence was manipulated or altered.

Some things haven't changed. A crime scene investigator's word is still his or her bond. Without integrity and trust, the whole enterprise can't function.

Prior to releasing the crime scene, law enforcement personnel need to hold a debriefing to ensure their work is complete and that everyone is clear about what needs to happen next. In a complex case, this may require several meetings and interagency discussions involving a detailed review of all of the evidence collected as well as its potential importance. Once this has been done and a final walk-through conducted, the investigator in charge needs to marshal all of the documentation into a case file that will allow for independent review of the work conducted.

Securing the crime scene takes on an even greater urgency than usual: the perimeter of American Media Inc. in Boca Raton is taped and the company is closed when one employee dies and another is found to be infected with anthrax, October 8, 2001.

Crime Scene Search: JonBenét Ramsey Murder, 1996

A rapidly unfolding situation in Boulder underscores the need to secure and thoroughly search a crime scene.

The brutal sex murder of a six-year-old beauty queen in her own upscale Boulder, Colorado, home on Christmas night in 1996 was so shocking that media outlets covered it as zealously as the Lindbergh kidnapping or the Simpson-Goldman slaying. To this day, JonBenét Ramsey's death remains unsolved, despite a long trail of strange clues found at the crime scene.

The mystery started at 5:51 a.m. on December 26, 1996, when Patsy Ramsey called 911 from 755 15th Street to report her daughter had been kidnapped. Boulder police quickly arrived at the large, luxury home to find Mrs. Ramsey, a housewife and former West Virginia beauty queen, and her husband, John Ramsey, 53, the millionaire president of a Colorado-based high-tech company named Access Graphics, upset by the disappearance and apparent kidnapping for ransom of their youngest daughter, little JonBenét.

Neighbors peeked out their windows, wondering what had brought a police car to the home.

The first responder, Officer Rick French, immediately searched the house's exterior for some sign of forced entry, but found none. Inside, Mrs. Ramsey showed French a three-page, handwritten ransom note she said she had found on the stairs after she had risen earlier that morning. French read the letter that someone had jotted with a felt-tipped pen on ruled white paper. Its contents seemed to indicate that a person or persons unknown had taken the girl and now were threatening to kill her unless they received a cash payment. The note said:

> Mr. Ramsey,
>
> Listen Carefully! We are a group of individuals that represent a small foreign faction. We do respect your bussiness [sic] but not the country that it serves. At this time we have your daughter in our posession [sic]. She is safe and unharmed and if you want her to see 1997, you must follow our instructions to the letter.
>
> You will withdraw $118,000.00 from your account. $100,000 will be in $100 bills and the remaining $18,000 in $20 bills. Make sure that you bring an adequate size attache to the bank. When you get home you will put the money in a brown paper bag. I will call you between 8 and 10 am tomorrow to instruct you on delivery. The delivery will be exhausting so I advise you to be rested. If we monitor you getting the money early, we might call you early to arrange an earlier delivery of the money and hence a [sic] earlier delivery pick-up of your daughter.
>
> Any deviation of my instructions will result in the immediate execution of your daughter. You will also be denied her remains for proper burial. The two gentlemen watching over

The Ramseys in May 1, 1997, four months after the murder. Four years later, after the grand jury failed to indict, Pat Ramsey told *USA Today*: "If you [prosecutors] think I did it, let's have a trial and get it over with."

your daughter do not particularly like you so I advise you not to provoke them. Speaking to anyone about your situation, such as police, F.B.I., etc., will result in your daughter being beheaded. If we catch you talking to a stray dog, she dies. If you alert bank authorities, she dies. If the money is in any way marked or tampered with, she dies. You will be scanned for electronic devices and if any are found, she dies. You can try to deceive us but be warned that we are familiar with law enforcement countermeasures and tactics. You stand a 99% chance of killing your daughter if you try to outsmart us. Follow our instructions and you stand a 100% chance of getting her back. You and your family are under constant scrutiny as well as the authorities. Don't try to grow a brain John. You are not the only fat cat around so don't think that killing will be difficult. Don't underestimate us John. Use that good southern common sense of yours. It is up to you now John!

Victory!

S.B.T.C

John Ramsey said that as soon as his wife showed him the note, he told her to call 911, even though the letter warned against notifying law enforcement. He said he then commenced a search and confirmed that she was gone from her bed. He also checked on his son Burke, aged nine, and found him still asleep in his second-floor bedroom near his sister's.

Soon a second patrolman responded, then four detectives and additional uniformed police arrived, along with an FBI agent from the Boulder office. Friends and neighbors of the Ramseys also appeared. Mrs. Ramsey was distraught and her husband said he regretted he hadn't set the burglar alarm. Amid the chaos some of the officers and friends searched the home for signs of the girl or other clues. Such actions would prove disastrous for the crime scene investigation.

Boulder was considered a safe community and the Ramseys were wealthy and respectable citizens. The local police had no experience with kidnappings and very little with homicide.

By 7:30 a.m. John Ramsey had obtained the $118,000 called for in the ransom note and he prepared to meet the demands at the same time he was working with the police. Based on what the detectives had found so far, they began to treat the case as a kidnapping. They arranged for phone taps and tape recorders to be set up, in the event that the kidnapper called. Other officers dusted the girl's bedroom for fingerprints and looked for other clues. Parts of the house still swarmed with the Ramseys, their friends, and police.

When the kidnappers didn't call at 10 o'clock, the Ramseys became more agitated. At the detectives' request, the couple provided handwriting samples and examples of shopping lists and other items that Patsy Ramsey had written, some of it scribbled on white ruled paper that was similar to the kind used for the ransom note. Just before 10:30 a.m. the lead detective ordered the girl's bedroom sealed and he set about trying to clear the house of nonessential persons. The Ramseys were told to remain in a designated area, but the chaos continued amid mounting fears about the girl's safety. With many of the police ordered away from the scene, the remaining officers had more trouble maintaining order. At 10:30 a.m., more FBI agents arrived and the police commenced a more thorough search of the 7,000-square-foot house and the grounds. Some of the agents scrutinized the ransom note and found many of its aspects highly unusual. Other officers fanned out to question some of Ramsey's business associates and neighbors.

Shortly before 1 p.m., the female police officer in charge of keeping the Ramseys in check suggested that John Ramsey and two of his friends accompany her on another search of the house to see if they could detect anything else missing. Ramsey led a business companion through the basement, checking various areas, and he entered a tiny windowless room called the "wine room," where his wife had recently hidden some Christmas presents. At the threshhold, he suddenly stopped and gasped.

"Oh my God, oh my God," he cried.

His daughter's prone body was lying on her back on the floor, partly hidden by a blanket. Ramsey rushed over and saw her neck and wrists bound tight with a cord, and duct tape over her mouth. He ripped off the tape and proceeded to try to loosen the cord as his companion knelt and touched the girl's flesh. It was cold and lifeless. Her body was rigid.

Ramsey scooped her into his arms and ran from the room, heading through the hall and up the stairs, yelling that he had found her. Before others could intervene, he placed her body next to the family's Christmas tree, near the unwrapped presents.

She was clad in a snow-white nightgown. A police officer noticed that the girl's hands were extended above her head from rigor mortis and a crude ligature had been fastened about her neck, another to her wrist. The little girl was dead, apparently garroted. Her lips were blue.

What had started as a kidnapping case was now a murder—a very high-profile murder.

THE AUTOPSY REPORT RAISES SUSPICIONS

Upon examining the child's body, Medical Examiner John Meyer determined that the girl had died of ligature strangulation and cranial damage—her skull had been fractured in a way that suggested she may have struck her head against a bathtub or toilet. The choking had been carried

The ransom note undergoes analysis, January 10, 1997. Handwriting analysis, if anything, eliminated the parents. The crime scene was not sealed off and that vitiated other forensic evidence.

out using a cord that was tightened by part of a paint-brush handle. Her body displayed other abrasions and signs of trauma and chronic inflammation of the vaginal tract, indicating that she may have been sexually abused on more than one occasion, but this has been questioned by experts such as Dr. Thomas Henry, the Denver medical examiner. As Meyer dictated his report into a tape recorder, a coroner's assistant and a police detective shot photographs of the body at every stage of the autopsy.

AUTOPSY REPORT
NAME: Ramsey, JonBenet AUTOPSY
NO. 96A-155
DOB: 08/06/90 DEATH D/T:
12/26/96 @ 1323
AGE: 6Y AUTOPSY D/T: 12/27/96 @
0815

SEX: F ID NO: 137712

FINAL DIAGNOSIS:
I. Ligature strangulation
    A. Circumferential ligature with associated ligature furrow of neck
    B. Abrasions and petechial hemorrhages, neck
    C. Petechial hemorrhages, conjunctival surfaces of eyes and skin of face
II. Craniocerebral injuries
    A. Scalp contusion
    B. Linear, comminuted fracture of right side of skull
    C. Linear pattern of contusions of right cerebral hemisphere
    D. Subarachnoid and subdural hemorrhage
    E. Small contusions, tips of temporal lobes
III. Abrasion of right cheek
IV. Abrasion/contusion, posterior right shoulder
V. Abrasions of left lower back and posterior left lower leg
VI. Abrasion and vancular congestion of vaginal mucosa
VII. Ligature of right wrist

Case Study: JonBenét Ramsey

CLINOCOPATHOLIGICAL CORRELATION:

Cause of death of this six year old female is asphyxia by strangulation associated with craniocerebral trauma.

John E. Meyer M.D., Pathologist, jn/12/27/96

The body of this six year old female was first seen by me after I was called to an address identified as 755 - 15th street in Boulder, Colorado, on 12/26/96. I arrived at the scene approximately 8 PM on 12/26 and entered the house where the decedent's body was located at approximately 8:20 PM. A brief examination of the body disclosed a ligature around the neck and a ligature around the right wrist. Also noted was a small area of abrasion or contusion below the right ear on the lateral aspect of the right cheek. A prominent dried abrasion was present on the lower left neck. After examining the body, I left the residence at approximately 8:30 PM.

EXTERNAL EVIDENCE OF INJURY: Located just below the right ear at the right angle of the mandible, 1.5 inches below the right external auditory canal is a 3/8 x 1/4 inch area of rust colored abrasion. In the lateral aspect of the left lower eyelid on the inner conjunctival surface is a 1 mm in maximum dimension petechial hemorrhage. Very fine, less than 1 mm petechial hemorrhages are present on the skin of the upper eyelids bilaterally as well as on the lateral left cheek. On everything the left upper eyelid there are much smaller, less than 1 mm petechial hemorrhages located on the conjunctival surface. Possible petechial hemorrhages are also seen on the conjunctival surfaces of the right upper and lower eyelids, but liver mortis on this side of the face makes definite identification difficult.

A deep ligature furrow encircles the entire neck. The width of the furrow varies from one-eight of an inch to five/sixteenths of an inch and is horizontal in orientation, with little upward deviation. The skin of the anterior neck above and below the ligature furrow contains areas of petechial hemorrhage and abrasion encompassing an area measuring approximately 3 x 2 inches. The ligature furrow crosses the anterior midline of the neck just below the laryngeal prominence, approximately at the level of the cricoid cartilage. It is almost completely horizontal with slight upward deviation from the horizontal towards the back of the neck. The midline of the furrow mark on the anterior neck is 8 inches below the top of the head. The midline of the furrow mark on the posterior neck is 6.75 inches below the top of the head.

The area of abrasion and petechial hemorrhage of the skin of the anterior neck includes on the lower left neck, just to the left of the midline, a roughly triangular, parchment-like rust colored abrasion which measures 1.5 inches in length with a maximum width of 0.75 inches. This roughly triangular shaped abrasion is obliquely oriented with the apex superior and lateral. The remainder of the abrasions and petechial hemorrhages of the skin above and below the anterior projection of the ligature furrow are nonpatterned, purple to rust colored, and present in the midline, right, and left areas of the anterior neck. The skin just above the ligature furrow along the right side of the neck contains petechial hemorrhage composed of multiple confluent very small petechial hemorrhages as well as several larger petechial hemorrhages measuring up to one-sixteenth and one-eight of an inch in maximum dimension. Similar smaller petechial hemorrhages are present on the skin below the ligature furrow on the left lateral aspect of the neck. Located on the right side of the chin is a three-sixteenths by one-eight of an inch area of superficial abrasion. On the posterior aspect of the right shoulder is a poorly demarcated, very superficial focus of abrasion/contusion which is pale purple in color and measures up to three-quarters by one-half inch in maximum dimension. Several linear aggregates of petechial hemorrhages are present in the anterior left shoulder just above deltopectoral groove. These measure up to one inch in length by one-sixteenth to one-eight of an inch in width. On the left lateral aspect of the lower back, approximately sixteen and one-quarter inches and seventeen and one-half inches below the level of the top of the head are two dried rust colored to slightly purple abrasions. The more superior of the two measures one-eight by one-sixteenth of an inch and the more inferior measures three-sixteenths by one-eight of an inch. There is no surrounding contusion identified. On the posterior aspect of the left lower leg, almost in the midline, approximately 4 inches above the level of the heel are two small scratch-like abrasions which are dried and rust colored. They measure one-sixteenth by less than one-sixteenth of an inch and one-eight by less than one-sixteenth of an inch respectively.

On the anterior aspect of the perineum, along the edges of closure of the labia majora, is a small amount of dried blood. A similar small amount of dried and semifluid blood is present on the skin of the fourchette and in the vestibule. Inside the vestibule of the vagina and along the distal vaginal wall is reddish hyperemia. This hyperemia is circumferential and perhaps more noticeable on the right side and posteriorly. The hyperemia also

Spectral analysis of the note revealed nothing. DNA forensics is an excellent tool for eliminating suspects, but of course in this case it could not eliminate the Ramseys because they had handled the note.

appears to extend just inside the vaginal orifice. A 1 cm red-purple area of abrasion is located on the right posterolateral area of the 1 x 1 cm hymeneal orifice. The hymen itself is represented by a rim of mucosal tissue extending clockwise between the 2 and 10:00 positions. The area of abrasion is present at approximately the 7:00 position and appears to involve the hymen and distal right lateral vaginal wall and possibly the area anterior to the hymen. On the right labia majora is a very faint area of violent discoloration measuring approximately one inch by three-eighths of an inch. Incision into the underlying subcutaneous tissue discloses no hemorrhage. A minimal amount of semiliquid thin watery red fluid is present in the vaginal vault. No recent or remote anal or other perineal trauma is identified.

REMAINDER OF EXTERNAL EXAMINATION: The unembalmed, well developed and well nourished Caucasian female body measures 47 inches in length and weight an estimated 45 pounds . . .

INTERNAL EXAM: . . .
G.I. Tract: . . . The proximal portion of the small intestine contains fragmented pieces of yellow to light green-tan apparent vegetable or fruit material which may represent fragments of pineapple ...
Skull and Brain: Upon reflection of the scalp there is found to be an extensive area of scalp hemorrhage along the right temporoparietal area extending from the orbital ridge, posteriorly all the way to the occipital area. This encompasses an area measuring approximately 7 x 4 inches. This grossly appears to be fresh hemorrhage with no evidence of organization. At the superior extension of this area of hemorrhage is a linear to comminuted skull fracture which extends from the right occipital to posteroparietal area forward to the right frontal area across the parietal portion of the skull. The posteroparietal area of this fracture is a roughly rectangular shaped displaced fragment of skull measuring one and three-quarters by one-half inch. The hemorrhage and the fracture extend posteriorly just past the midline of the occipital area of the skull. This fracture measures approximately 8.5 inches in length . . . On the right cerebral hemisphere underlying the previously mentioned linear skull fracture is an extensive linear area of purple contusion extending from the right frontal area, posteriorly along the lateral aspect of the parietal region and into the occipital area. This area of contusion measures 8 inches in length with a width of up to 1.75 inches. At the tip of the right temporal lobe is a one-quarter by one quarter inch similar appearing purple contusion . . .

EVIDENCE: Items turned over to the Boulder Police Department as evidence include: Fibers and hair from clothing and body surfaces; ligatures; clothing; vaginal swabs and smears; rectal swabs and smears; oral swabs and smears; paper bags from hands, fingernail clippings, jewelry, paper bags from feet; white body bag; sample of head hear, eyelashes and eyebrows; swabs from right and left thighs and right cheek; red top and purple top tubes of blood.

As Meyer announced his findings at a news conference, attention suddenly shifted to the Ramseys. The bizarre tableau left behind prompted some experts to wonder if the crime scene was staged to divert suspicion.

A horde of national news media hounded police and prosecutors, the Ramseys and anyone associated with the case, trying to get any information they could about the girl's bizarre murder. Nationally known lawyers and forensics experts were also showcased in cable TV news programs to opine about the case.

The Ramseys themselves were quickly targeted. John and Patsy Ramsey each hired top defense lawyers to represent them, even though they had not been charged with a crime. They also brought in their own public relations consultants and declined to submit to polygraph examinations. Appearing with their family doctor, they disputed any suggestion that their daughter had been chronically sexually abused. They also hired their own private investigator. Some pundits questioned why they appeared more concerned about protecting themselves than helping law enforcement authorities solve their daughter's murder.

Boulder police were portrayed in the media as having botched the investigation and the DA was criticized as weak. Critics charged that the police had failed to seal off the crime scene promptly, thereby allowing crucial evidence to become overlooked and contaminated. They had also failed to search the scene thoroughly before jumping to conclusions. The fact that they had allowed the body to be discovered—and then moved—by John Ramsey, after he had altered the tape and other evidence, was particularly assailed.

Leaks from the investigation disclosed that police had also found a "practice" ransom note in the house, indicating that the killer or an accomplice had taken additional time to formulate a note before the final version was completed. Both notes had apparently been composed at the crime scene, using materials from the crime scene. Handwriting experts failed to come to any firm conclusions about whether the notes may have been written by one of the Ramseys.

Some investigators also commented on the suspicious nature of the ransom note itself. They pointed out that it seemed to reflect an intimate knowledge of some aspects of John Ramsey's private life.

The odd figure of $118,000, for example, matched the amount of his recent company bonus and the letter also contained a reference to the Navy base in the Philippines where he was once stationed. The phrase, "Use that good southern common sense of yours," indicated that the writer knew the Ramseys were originally from Atlanta, not Boulder. Some experts thought a professional kidnapper would have demanded more money than $118,000.

Kidnapping experts pointed out that ransom notes are usually written ahead of time and typed to conceal authorship. Everything about the note seemed to suggest someone who was criminally unsophisticated.

Colorado profilers suggested that the placement of JonBenét's body was telling, because an intruder would not have wanted to leave the evidence in the house, whereas a parent would have been psychologically inclined not to hide the child's body outside. They also concluded that the fact that her body was partially covered indicated remorse. Likewise, the nature of the garrote was considered significant. It was tied from behind, possibly to prevent eye contact between the killer and the victim.

Some detectives concluded that the girl probably had been accidentally struck on the head and the other injuries had been staged to cover up the assault and divert attention elsewhere. They speculated that the girl may have wet her bed, upsetting her mother. Patsy Ramsey may have lost her temper and inadvertently slammed the child into a tub or sink, thereby fracturing her skull. Everything else from that point, they theorized, was part of the coverup.

Other investigators sided with the Ramseys, claiming that an intruder or intruders had entered the home and carried out the diabolical acts after reading newspaper accounts about Ramsey's business success and his daughter's beauty-pageant charms. The controversy became so heated that two investigators involved in the case resigned in protest over the way the case was being handled. After some of the grisly crime scene photos ended up in a supermarket tabloid, a police photo-processor was charged with felony theft, evidence tampering, and lesser offenses.

In the end, the Boulder District Attorney failed to convince a grand jury to indict any suspects, leaving the case still open. The Ramsey murder remains one of the nation's leading unsolved mysteries, a textbook example of what can go wrong in crime scene investigation.

THE SCIENTISTS

Werner Spitz, M.D., forensic pathologist—Medical Examiner, Spitz has investigated the deaths of JFK, Martin Luther King, and Mary Jo Kopechne, and served as an expert witness in numerous high-profile cases—the California Night Stalker, the Preppy Murder Trial in New York, the wrongful death suit against O.J. Simpson, and the Ramsey case in Boulder. He is the author of *Medicolegal Investigation of Death*, the leading textbook in forensic pathology.

Crime Scene Reconstruction

How forensics experts try to use the physical evidence and other input to make the crime come alive. Until it is analyzed and used to support a conclusion, forensic evidence collected from the crime scene has very little value.

Detective work has long entailed efforts to develop hypotheses in an effort to solve crimes. Since the 1990s, however, a new science known as crime scene reconstruction has developed.

Crime scene reconstruction is the process of analyzing the physical evidence taken from the crime scene, studying its location, position, and role in the crime, along with the laboratory examination of the physical evidence, knowledge of the likely behaviors of participants under known conditions, and an application of common sense inductive and deductive reasoning in an effort to determine what happened, what was the most likely sequence of events.

Collecting and studying physical evidence at the crime scene forms only the initial step in the reconstruction process. First the investigator must be able to spot it and recognize, for example, that the discoloration represents a bite mark or that the blood-spatter pattern found at neck-level on a wall indicates an arterial gush. Only after this evidence is documented and a permanent record is established, can the investigator then seek to identify the evidence by comparing it with other like bite marks or blood patterns.

Efforts can then be made to determine its uniqueness—for example, to subject the bloodstain to DNA analysis, or to use the latent fingerprint found at the scene to make a fingerprint identification of the individual.

The reconstructionist uses this information to try to interpret what happened at the crime scene. This is done by considering different possible interpretations or explanations before forming a hypothesis. In keeping with the scientific method, this hypothesis is then tested to confirm or disprove it until finally a theory of the crime emerges. "The scientific method" is not just a phrase: it follows that sequence precisely. Observe, form a theory that is consistent with what has been observed; use the theory to make predictions; test those predictions through experiments or further observations and modify the theory, or hypothesis, in the light of the results. Continue to test/observe and modify the theory until there is no discrepancy.

Seasoned investigators are aided by their knowledge of pattern evidence from the crime scene: blood-stain patterns, tire mark patterns, injury or wound patterns, clothing or furniture damage positions, and so on. Each potential pattern must be carefully documented and studied. What does the wound pattern reveal about the sequence of the attack? How much force was used to stab the victim? Crime scene reconstruction is a dynamic process that puts physical evidence to work to help tell the story of what happened.

Dr. Henry C. Lee

What makes the amazing Dr. Henry Lee, the "King of Crime Scene Investigation." Dr. Henry Chang-Yu Lee is generally regarded as the greatest crime scene expert in American history, a veteran of more than 45 years in policing, a master biochemist, and a prolific writer about scientific crime-solving. He's also someone who in person often comes across as a Chinese-American Sherlock Holmes; a sage who embodies wit, determination, and patience. His personal story is more amazing than any fictional detective's.

Born into a Buddhist family in Rugao city, Jiangsu province, China, on November 22, 1938, Lee's interest in violent crime started at age four when his father was murdered by the

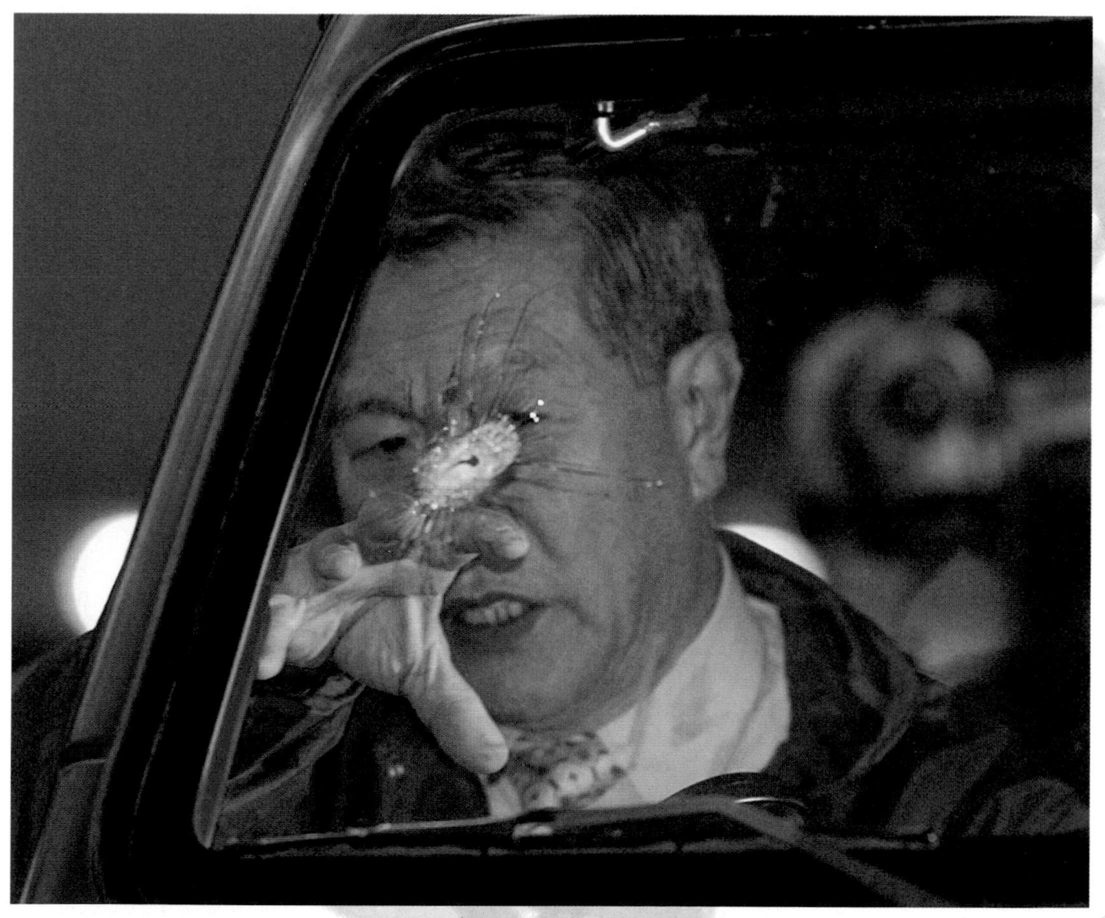

Henry Lee inspects a bullet hole in the windscreen of the car carrying Taiwan President Chen Shui-bian and Vice President Annette Lu on March 19, 2004, the day before they were narrowly re-elected. Both were hit by bullets as they drove in an open-top Jeep through the streets of Tainan, but were not seriously injured. Lee's 130-page report concluded that this was not a serious assassination attempt because "a more powerful weapon than a home-made pistol would have been used."

Communists, eventually forcing his mother to flee with him and 12 other older siblings to Formosa to escape Mao's revolution. He grew up fascinated by biology and developed a penchant for dissecting animals in his school laboratory.

After graduating first in his class from the Central Police College in Taiwan in 1960, he rose to the rank of captain by age 22, the youngest ever, and married a Malaysian woman he had arrested on a student visa violation. They remain married to this day.

In the mid-1960s Lee moved to the United States with only $50 in his pocket and practically no proficiency in the English language, but he soon managed to make his way by virtue of his own intelligence and hard work. He and his family became Americans. In 1972 he earned a B.S. degree in forensic science from John Jay College of Criminal Justice, and later went on to receive an M.S. and Ph.D. in biochemistry from New York University in 1974 and 1975. Since then he has also accumulated several honorary advanced degrees.

In 1979 Governor Ella T. Grasso of Connecticut tapped him to direct the State Police Crime Laboratory and oversee major crime scenes, but warned him, "We can't afford to pay much." She wasn't kidding: his first crime lab consisted of a bathroom with one microscope. But under Lee's hard-working leadership, the facility grew into a modern forensic laboratory with an educational program that has made it ranked as one of the best in the world.

After ten years, by working seven days a week, often 16 hours a day, and being always on call, Lee had assembled more than $10 million worth of laboratory equipment and boosted the annual budget to more than $2 million, making Connecticut the only state in New England to already be typing DNA for blood, semen, hair roots, and cheek cells. He had also harnessed computer power to vastly increase the lab's analytical capabilities.

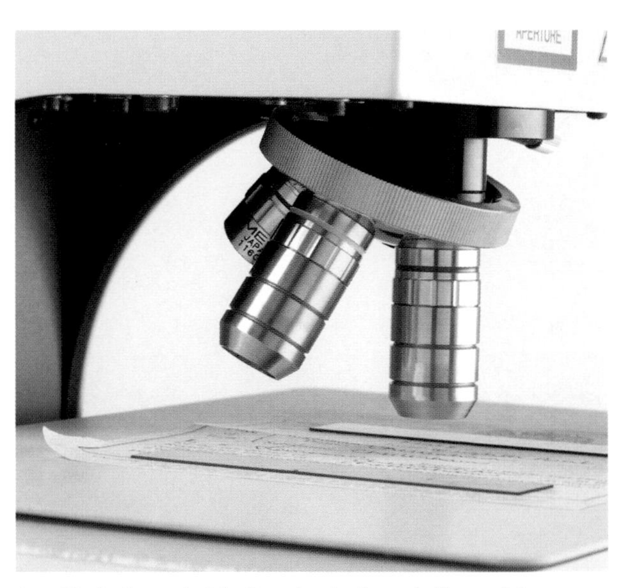

Arguably the first tool of the forensic scientist, and still one of the most important, the microscope. When Dr. Lee first opened for business as head of the new Connecticut State Police Crime Laboratory, this was about all he had.

Once a day, at 5:30 p.m., he ritualistically "blanked his mind" as a way of coping with the constant stress and the trauma that perpetually investigating murder, rape, and other crimes entailed.

Lee served as Connecticut's Chief Criminalist, Director of the State Police Forensic Science Laboratory in Meriden, and Commissioner of the Department of Public Safety before becoming chief emeritus of the

Connecticut Department of Public Safety Division of Scientific Services. He also has spent much of his time teaching and serving as an extremely active forensic consultant for both the prosecution and the defense. As a distinguished professor, he teaches a field course in the Forensic Science Program at the University of New Haven, and edits as many as seven academic journals.

Lee has been a consultant for more than 900 law enforcement agencies and investigated more than 7,000 criminal cases. He's appeared in court in many high-profile death cases including O.J. Simpson, Vince Foster, and Elizabeth Smart, authored 20 popular books, and often appeared on *Larry King Live* and other television shows, even hosting one of his own popular programs.

Renowned as a top expert in many aspects of forensic science, including fingerprint analysis, blood spatter analysis, and DNA trace analysis, since the 1980s, through his writings, his development of new techniques, his training and teaching, and his legendary, hands-on involvement in baffling major criminal investigations, Lee has helped to revolutionize the field of crime scene investigation by bolstering a relentlessly professional scientific approach to crime solving. He once told an interviewer:

> "A good crime solver needs to use basic common sense plus have scientific knowledge and understand human nature."

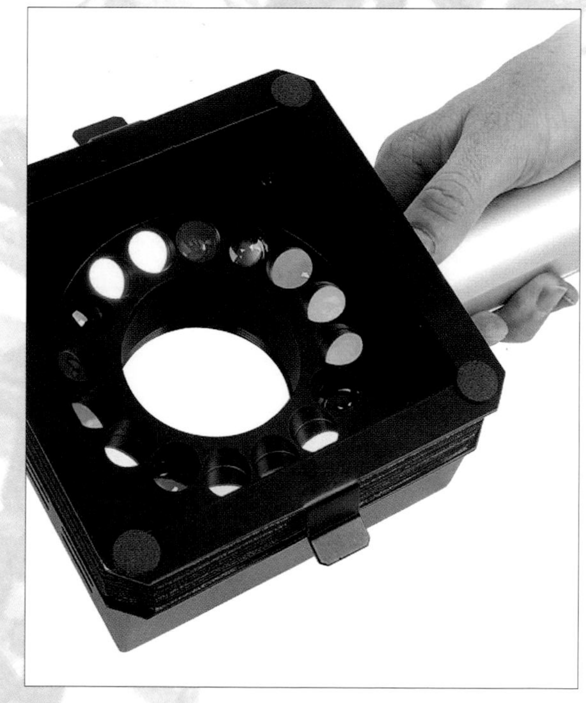

The Woodchipper case was a classic example of minute and painstaking examination of a crime scene. Forensic lamps such as these were used to find hairs and blood spots that matched those of the victim. The 4 x 4 high-intensity LED light on the right is used for forensic photography, providing a higher output than conventional filtered xenon or metal halide light sources.

Lee first burst into the national spotlight in the 1980s due to his work in the grisly Woodchipper case, which entered the annals of forensic science as an especially noteworthy episode that helped redefine the meaning of *corpus delicti*, or the body of the crime.

As Lee himself would later write in *Henry Lee's Crime Scene Handbook*, the term *corpus delicti* doesn't simply signify the corpse in a death investigation, as the Latin definition would indicate, but it "is the determination of the essential facts that will show that a crime has occurred."

In the Woodchipper case, Lee and other members of his forensic team managed for the first time in Connecticut history to secure a murder conviction without a corpse, demonstrating in the process an uncanny ability to marshal several different forensic techniques in collaborative pursuit of a murderer who thought he had pulled off the perfect crime.

The case started in November 1986 with the disappearance of Helle Crafts, a strikingly beautiful Danish-born flight stewardess for Pan American Airlines who lived with her husband in upscale Newtown, Connecticut. Richard Crafts worked as a commercial airline pilot and had previously flown missions on Air America for the Central Intelligence Agency; he was a man of adventure and intrigue. Crafts went to the police saying his wife had gone away after an argument and never returned, prompting him to file a missing persons report. But friends of the missing woman alerted investigators that she planned to divorce her husband and had said she was afraid for her life.

Although police began to suspect foul play, they had no evidence that a crime had been committed. But their suspicions increased when a witness came forward to report that he had been out plowing snow at 4 a.m. at the time of the disappearance when he had observed Richard Crafts acting suspiciously. He said he had spotted Crafts in a remote location overlooking the Housatonic River. And the strangest thing was that the flight skipper was frantically operating a huge woodchipper at the height of a major snowstorm.

Detectives found out that, immediately before Helle Crafts was reported missing, her estranged husband had rented an Asplundh Badger Brush Bandit 100 model woodchipper, an unusually large model for someone doing non-commercial work. During the same frenzied period, Richard Crafts had also gotten a new truck, picked up a large Westinghouse Chest Freezer, using an alias, and traveled outside his immediate area to purchase several other noteworthy items such as rubber gloves and a flathead shovel. Further investigation revealed that Crafts had been cheating on his wife and exhibiting some bizarre behavior involving firearms. To their horror, police began to suspect that he may have killed his mate, dismembered her body with a chain saw, shredded the corpse through the woodchipper, and emptied the pieces into the river.

After routine searches turned up nothing, and the police still lacked any body or blood to prove that a murder had taken place, Dr. Lee set out to comb the possible crime scenes for biological traces or other physical evidence. When ordinary inspection failed to turn up any brown stains, Lee and his team methodically scoured the couple's home using more sophisticated methods. At several suspicious locations, Lee sprayed luminol, a chemical mixture that can help

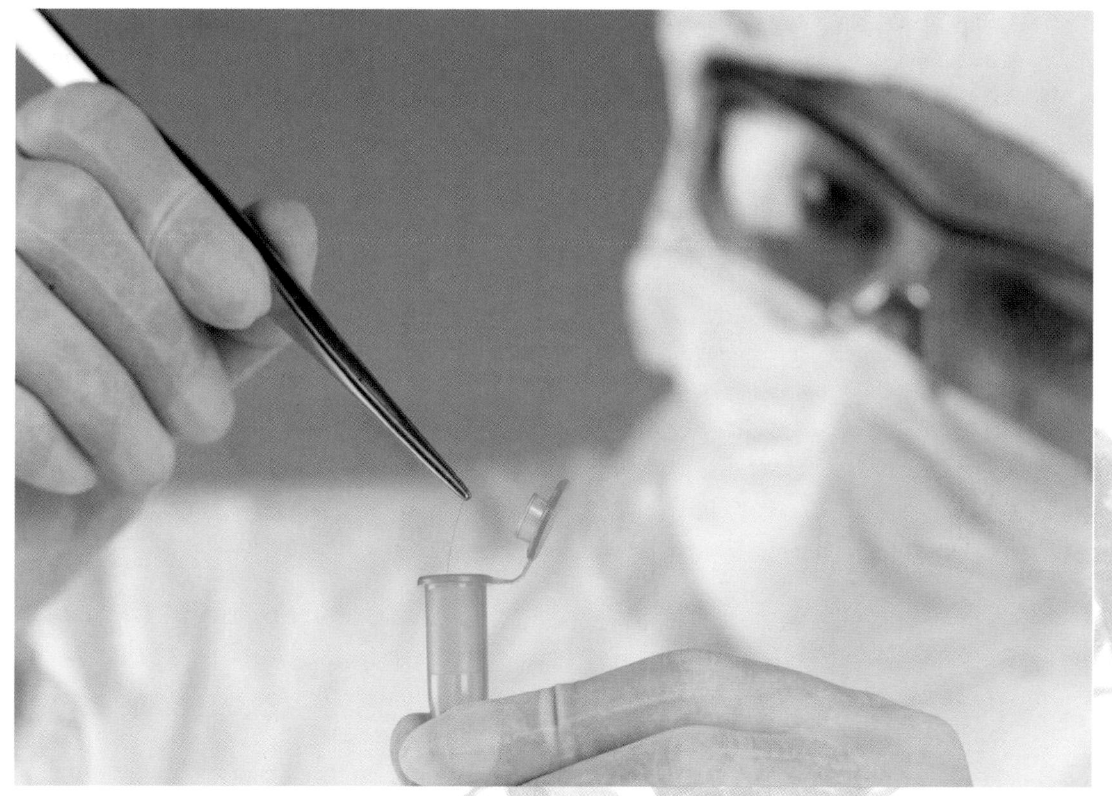

A single hair, either from a suspect, a crime scene, or a victim, is collected and catalogued. Hairs can actually be analyzed to find evidence of toxic intake and of drugs, indicating, by measurement of the hair growth and the position of the toxic residue along the hair shaft, when such drugs were taken, or poisons ingested.

investigators detect stains, especially the heme portion of hemoglobin in red blood cells, that are otherwise invisible to the naked eye, and TMB (tetramethhylbenzidine), another blood enhancement reagent. This enabled him to discover several tiny stains of Type O blood, the same type as Helle Crafts'.

They also located the same woodchipper that Crafts had rented and minutely examined the river location where the snowplower had spied the suspect. Like any skilled crime scene investigator, Lee also noticed some items that were missing from the likely crime scene, such as the box spring of their bed, which caused him to hypothesize that Crafts may have committed the murder on the bed and later disposed of that part due to the bloodstains.

After many long hours of painstaking search the team assembled a catalogue of telltale evidence, including human blood spots, more than sixty tiny chips of human bone, 2,660 strands of blond hair resembling that of Helle Craft, and part of a finger containing fingernail polish that matched a cosmetic found in her home makeup kit.

Eventually, the evidence persuaded a grand jury to indict Richard Crafts for the murder of his wife. Newspaper tabloids plastered their pages with gruesome details about the alleged killing, one of them calling it, "DIVORCE, CONNECTICUT STYLE."

At the first trial held in New London, Dr. Lowell Levine, the famous forensic dentist, positively identified a tooth fragment from the scene as having belonged to Helle Crafts' lower-left bicuspid, but the contest ended with a hung jury.

Then a second trial, held in Stamford in March 1989, ultimately resulted in Crafts' murder conviction. He was sentenced to 50 years in prison—and Lee's reputation was carved out, not just in the media, but also in police and legal circles.

Under Lee's expert guidance, the celebrated Woodchipper Case had brought together odontology, serology, fabric and hair examination, pathology, time-line analysis, weather evidence, credit card tracing, and other techniques to determine the essential facts of the crime—the *corpus delicti*, or body of evidence.

Piece by piece, the Lee legend grew. Soon he became known as the "King of the Crime Scene," the best in the business. A few years later, Lee's crime scene reconstruction skills led to the capital conviction of Daniel Webb for the 1989 kidnapping, attempted rape, and murder of bank vice president Diane Gellenbeck in Hartford.

He also was called in to investigate the mysterious 1993 shooting death of Vincent Foster, White House counsel to President Bill Clinton, whose corpse had set off a wave of right-wing conspiracy theories. Lee concluded otherwise, noting for example that his examination of crime scene photos of Foster's glasses before they were washed by the medical examiner, had revealed tell-tale signs of blood spatter that proved the shooting was self-inflicted.

One of his most dramatic tests involved his role as a defense expert in the O.J. Simpson case—one of history's greatest landmarks in crime scene investigation.

Lee was hired by the defense after "Dream Team" attorney Robert Shapiro called from Los Angeles to say that EDTA, a synthetic preservative put in lab tubes to keep blood from coagulating, had been found in blood recovered from Simpson's socks on his bedroom floor. Lee agreed that such a finding implied that the blood may have been planted, and he agreed to join Michael Baden, Barry Scheck, Peter Neufeld, and other luminaries on Simpson's payroll—even though, as a forensic scientist, he insisted on maintaining his own independence and search for the truth.

The result was that Lee's crime scene investigation immediately led him to conclude, "Something's wrong." And his review of the official crime scene investigation demolished many crucial steps taken or not taken by the Los Angeles Police to find the evidence and preserve the sanctity of the evidence-custody chain. As a consequence, the field of crime scene investigation was forever changed.

To this day, Lee won't say publicly whether he believes O.J. Simpson was innocent or guilty. Sometimes he quotes Sherlock Holmes as saying: "Once you eliminate the impossible, whatever remains, no matter how improbable, must be the truth." But he argues that the Simpson case exerted a positive effect on law enforcement, by requiring the police to be more scientific and more professional in their crime scene investigation. He also took steps to take the $150,000 he made from the Simpson case and invest it in improvements at the University of New Haven's forensics program, for the betterment of the field.

During the Clinton presidency, Lee was also hired to test the stain on Monica Lewinsky's famous blue dress, and he once joked that Bill Clinton's DNA more resembled an undone zipper than a standard double helix.

Even as he has helped to make history in crime scene investigation, by meeting new challenges posed by DNA, he has also warned that even DNA, like fingerprints and other seemingly failsafe forms of evidence, can be fabricated or beaten as new technologies emerge. He likes to point out:

> "Let's say now you can clone somebody. All of a sudden we have 10 of you with identical DNA. Who the suspect?"

Sample tubes are color-coded for the collection of blood according to different hematological tests. Pink tubes contain EDTA, an anti-coagulant that will stop blood from clotting before tests can be carried out. What was EDTA doing in the socks found in O.J.'s bedroom? This anomaly was what persuaded Dr. Lee to become involved in the case.

Prints and Identification

Techniques for lifting and matching fingerprints have undergone vast improvements over the last seventy years—but the people and systems that use them are now vulnerable to challenge. The Oklahoma Bomber case did not involve fingerprints, but rather a series of "imprints," from a license plate, to an arrest record and chemicals on a vehicle, that all pointed at one suspect.

The introduction of fingerprint identification in the United States actually predated the Lindbergh case by 30 years, but its massive files and routine use took decades to develop. It has remained one of the staples of criminal investigation for more than seventy years.

Practically from the start, police have "dusted for prints" in an effort to find a criminal's calling card at the crime scene. In the old days, such as at the scene of the Lindbergh baby kidnapping, police used a camel hair brush to apply a standard black powder to a suspicious nonporous surface, then they blew off the excess powder in hopes of finding some that adhered to the moist, greasy lines left by human fingers. Such latent prints would then be photographed and lifted with a low-tack adhesive tape and mounted on an acetate sheet to be preserved as evidence.

Over the years, besides applying aluminum powder, the police developed other methods for more porous surfaces such as paper and cardboard. Techniques were also devised to develop prints with chemicals including iodine, ninhydrin, and silver nitrate. Some lab technicians also began to use physical developer (PD), a solution of silver and iron compounds, for porous surfaces that have been soaked in water. Certain enhancement techniques will only work for a short time before they eventually obliterate the prints, and therefore their use requires special skills and caution.

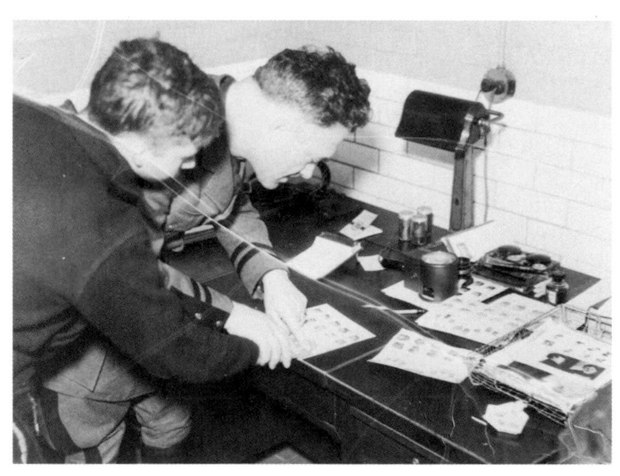

Fingerprinting in the early 1930s. In 1858 Sir William Herschel, a British Chief Magistrate in Jungipoor, India, began to insist that palm prints and later fingerprints were attached to contracts by the locals. It made it somehow more binding psychologically. Only later did he realize that the prints really could be used to prove identity, as each was unique.

By the start of the twenty-first century, forensic scientists had also added digital imaging, dye stains and fumes, thus bringing the list of fingerprinting methods to more than forty. Colored powders were developed to contrast with surface colors, and some powders or dyes became available that would glow under special light sources such as from a high-powered laser. Fuming with iodine vapor, for example, gives prints a brown color that fades rapidly.

Thus far, one of the most sensitive, time-consuming, and expensive lab techniques has been vacuum metal deposition (VMD). It entails putting the object to be tested into a pressurized vessel from which the air is pumped out and the chamber is filled with metal vapor: first gold, and then zinc. The metals condense on the ridge patterns, making the prints visible, although they may later need to be treated with superglue.

A crime scene investigator is required to obtain fingerprints from every person (included the deceased) who is known to have been at the crime scene. Some prints may also be removed from the victim's skin, but only with approval of the medical examiner and using accepted techniques.

A poster stressing the importance of having fingerprints on file as a form of identification during the Second World War. By 1946 the FBI had processed more than one hundred million fingerprint cards. Paper cards are still the main storage method today.

Once a crime scene fingerprint is photographed and preserved, police can use it to try to identify its source. Generally that means either comparing it with a specific suspect or entering it into a computerized database such as AFIS (Automated Fingerprint Identification System), to see if it can be matched with a known individual whose fingerprints are on file. The FBI's Integrated AFIS in Clarksburg, WV, holds more than 49 million computerized records for known criminals.

To make a match, a fingerprint examiner needs to look for clear points of comparison. A single fingertip may offer as many as three hundred of them. The more that can be found, then the more likely a match can be made. All it takes to rule out a match is one dissimilar point. Since 1972, the FBI had dramatically improved its ability to use computers to do the work it used to take an immense labor force to accomplish. Today's automated system scans and digitally encodes the prints into a geometric pattern. In less than a second, the computer can compare a set of ten prints against a half million prints stored in the database. The result produces a list of prints that most closely match the original. Then the technicians make a point-by-point visual comparison.

Fingerprinting, however, has lately come under fire. With the *Daubert* ruling of 1993, the

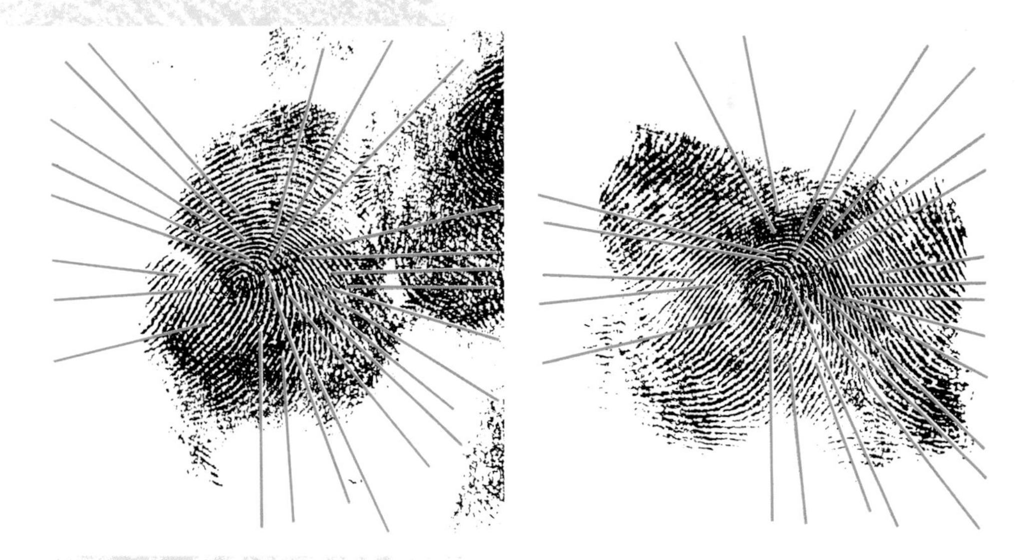

Locating the minutia—the characteristics that indicate the fingerprint at the scene of crime and of the suspect are the same. These are sometimes called Galton's Details, after Sir Francis Galton, who first classified prints by these points of reference in 1892. It is an old science, but by no means outdated. Fingerprints solve at least ten times as many unknown suspect crimes as DNA analysis worldwide.

United States Supreme Court established new guidelines for the acceptance of scientific evidence, including fingerprints, and this has subjected the science of fingerprinting and the proficiency of fingerprint examiners to unprecedented and long-overdue scrutiny. Sometimes such review has revealed serious weaknesses or even incompetence. In 1995, the Collaborative Testing Service administered a proficiency test sponsored by the leading professional body in the field, International Association for Identification. Of 156 examiners tested, only 68 (44 percent) correctly identified all seven latent fingerprints; 88 (56 percent) made at least one mistake, and 6 (4 percent) failed to identify any correctly. This added up to an overall error rate of 22 percent. In other words, nearly a quarter of the cases could have resulted in the wrong person being "positively identified" by the time-honored and "infallible" forensic method of fingerprinting.

Other forms of biometric identification have suddenly gained special appeal in the war on terrorism. (Biometrics is the science of identifying persons, based on unique physiological characteristics, by means of an automated system.) These techniques include palm-print systems, eye-identification systems, face recognition, speech recognition, ear-recognition systems, and the geometry of the hand. Retinal scans as used at airports today are not a new idea. It was recognized back in the 1930s that the pattern of blood vessels at the back of the eye was specific to individuals and did not change much over time: there was just no technology to make use of the fact. Some systems are being integrated with camera networks and criminal identification databases to enable police to identify subjects quickly. Many are already being used on a large scale by private companies, raising all kinds of legal and ethical issues.

AFIS: Richard Ramirez, the Night Stalker

A powerful new tool enables police to find the needle in the haystack and catch an infamous serial killer.

From June 1984 to August 1985, a rapist-killer stalked the night in Southern California, leaving comfortable suburban homes awash with blood and gore. Some of the victims were couples who had been asleep in their beds. Others were children as young as eight years old. The attacker broke in like a cat burglar, shot or hacked to death the male, then savagely raped and usually murdered the woman, leaving behind a trail of ransacked rooms, mutilated corpses, and Satanic pentagrams. One victim had her eyes gouged out. The viciousness of his attacks eventually earned him his own infamous nickname in the media. They called him the Night Stalker.

After one brutal rape of a 42-year-old mother in Burbank, the Night Stalker inexplicably left her behind as a witness. "I don't know why I'm letting you live," he told his victim. "I've killed people before. You don't believe me, but I have."

Based on her description, police began to seek a tall, thin, Hispanic-looking man with bad teeth and strong body odor. The LAPD released a photofit of the suspect, which was published in newspapers all over California and shown repeatedly on television. But they still didn't have a solid suspect.

By early August 1984, the LAPD had established a special Night Stalker task force with 200 detectives working around the clock. Investigators searched for prints and blood, canvassed neighborhoods, scanned databases, developed offender profiles, and even consulted with experts in devil worship and the occult. They papered the city with composite sketches and grilled their informants for clues.

Then the Night Stalker struck 400 miles away in San Francisco. After viciously attacking an elderly Chinese couple, one of them fatally, he left behind a lipstick-scrawled message to taunt his pursuers. A week later he struck again halfway between Los Angeles and San Diego, in

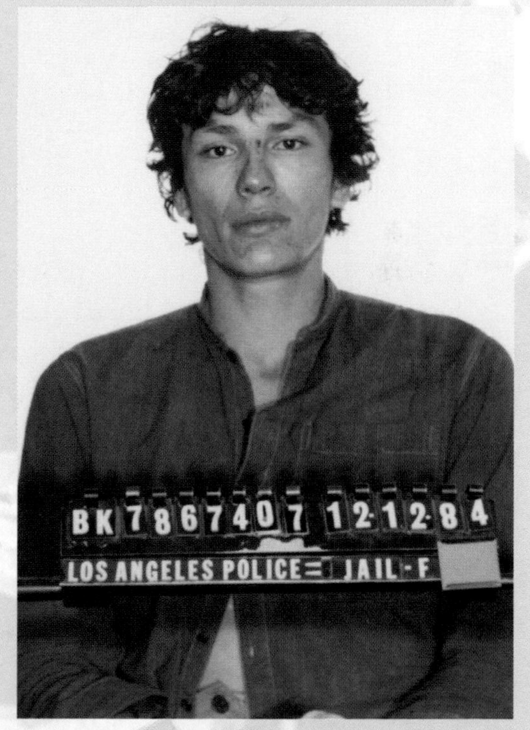

The lack of any similarity in Ramirez' victims was almost certainly a factor in the delay in capturing him. His first victim was a 79-year-old-woman, his next a six-year-old girl.

A portable cyanoacrylate fuming system, for use in small rooms and in vehicles to find latent fingerprints, with fans, a humidifier, and superglue evaporator. The Superglue test that caught Ramirez was a lot less "boutique."

Mission Viejo. Suddenly, all of California was on alert. The big breaks in the case came when one victim managed to catch a glimpse of the attacker's vehicle as he fled. About the same time, a sharp-eyed teenager, James Romero III, had also spied a suspicious car of the same general description prowling the neighborhood. Romero jotted down the Toyota station wagon's license plate number and gave it to police.

Police traced the number to a stolen car and put out an all-points-bulletin. Two days later they found the Toyota abandoned in a parking lot in the Rampart section. The cops kept the area under surveillance in hopes the killer would return. Meanwhile, the forensic team tried a special trick of the trade to comb it for evidence.

They placed a saucer of Superglue inside the car and shut the windows, hoping that the fumes from the glue would spread throughout the car and react with moisture to show up any latent prints. Sure enough, when a laser beam was passed over the car's interior, they discovered a single partial fingerprint and lifted it from the vehicle.

In the past, such evidence probably would never have amounted to anything because the Los Angeles Police had as many as 1.7 million fingerprint cards to search and compare by extremely laborious and time-consuming eyeballing. However, as luck would have it, the LAPD had just started installing a much-heralded new high-speed Automated Fingerprint Identification System (AFIS) that enabled analysts in Sacramento to compare more than 60,000 fingerprints per second, and some of the prints were already available for matching.

Almost instantly, the new system delivered the goods. Within a few minutes, AFIS registered a match – something it was calculated could have taken an experienced fingerprint examiner as long as 67 years to achieve. Bingo. The lead suspect was Richard Ramirez, aka Ricky, age 25, who had been arrested and convicted of car theft. As soon as his photograph and other information flashed over the screen, LAPD detectives working the case knew they had their man.

Police released his mugshot and description to the media, sending it screaming across TV screens all through the Los Angeles area and beyond.

At the time, Ramirez was returning from a cocaine-buying trip to Phoenix, unaware of the latest developments. But as he was waiting in line to buy his sugary breakfast at Tito's Liquor Store in East Los Angeles, he spied his own face looking up at him from the newspaper rack.

Other customers also began to recognize him as well. Bolting into the night, Ramirez frantically looked around for a car to steal, but residents and cops were closing in. Now he was the prey.

Ramirez sprinted through the barrio, racing through yards and jumping over fences, and finally tried to hijack somebody's car. But a mob of neighbors, some of them armed with metal bars, pulled him from the vehicle, and somebody shouted that he was the monster everybody was looking for. Police arrived just in time to save Ramirez from being torn apart or lynched.

Investigators found a murder weapon at the home of one of his friends and his sister gave up a piece of jewelry that had been taken from one of the victims. Ramirez was arrested under suspicion for 16 murders and 24 violent assaults. Before the case was done, investigators had interviewed nearly 1,600 witnesses.

His trial turned out to be one of the longest in U.S. history. Ramirez' chilling behavior in and out of the courtroom added to the case's grisly aura. But in the end, Ramirez was convicted and on November 7, 1989, he was sentenced to death in the gas chamber. "You don't understand me," he told the judge. "You are not capable of it. I am beyond your experience. I am beyond good and evil."

The forensic manufacturer's "Superfume" system not only saves time by locating fingerprints without painstaking dusting, but also because the prints are processed at the crime scene. The powerful steamer brings the humidity up to 80%HD prior to fuming.

Identification: Timothy McVeigh

In Oklahoma City, the deadliest terrorist bombing in U.S. history prompts rapid identification, following a trail of forensic clues.

On April 19, 1995, after the worst terrorist blast in U.S. history killed 168 or 169 people and injured more than 500 others at the Alfred P. Murrah Federal Building in Oklahoma City, many observers suspected Islamic fundamentalists, who already had attacked the World Trade Center in 1993, and assaulted U.S. embassies and military targets throughout the world. The Oklahoma bombing set off one of the most intense criminal investigations ever.

Even before the search had been completed for living or dead bodies, crime scene investigators focused much of their attention on the massive crater created by the explosion, and workers quickly proceeded to construct a wooden cover over the site to shield it from damage and hopefully preserve some of the evidence. Meanwhile, structural engineers continued to monitor the structural stability of the remaining wreckage, occasionally setting off blaring horns to warn the search teams of potential danger.

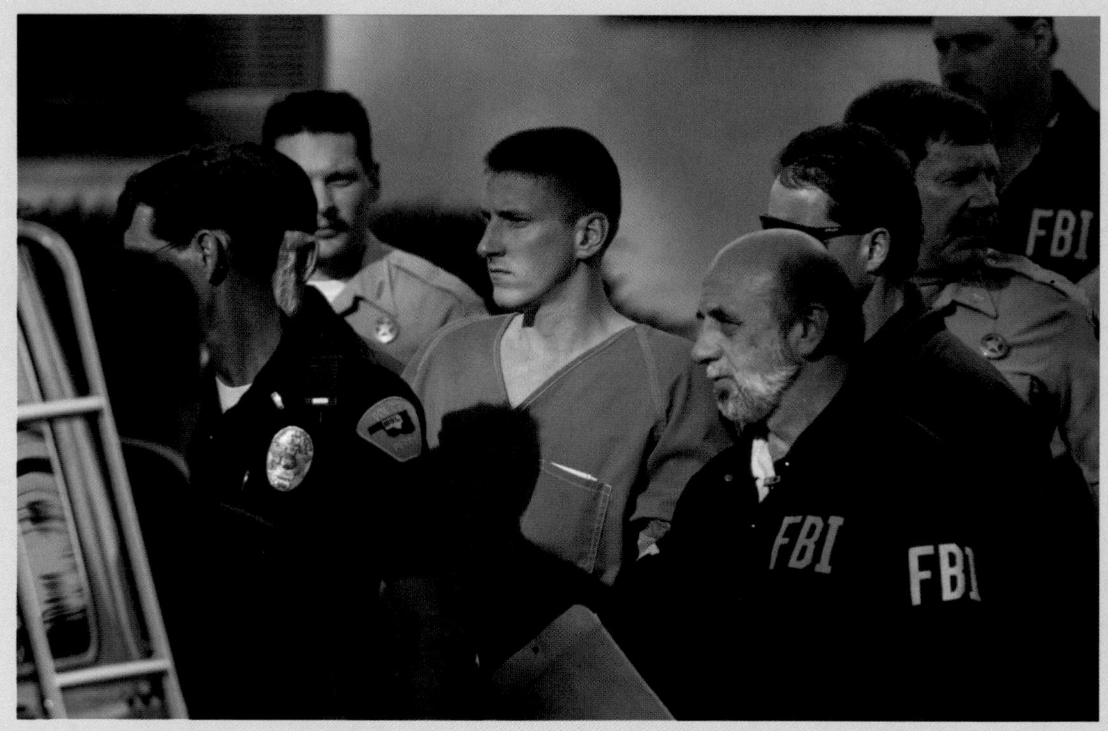

FBI agents and police officers escort Timothy McVeigh from the Noble County Court House, April 21, 1995. McVeigh was executed by lethal injection on June 11, 2001, for murdering 168 people.

Some of the damage assessment experts carried sophisticated EGIS explosives detectors capable of detecting traces of EDGN, NG, TNT, PETN, RDX, and other common high explosives. One of their goals was to find traces of the original explosive material that hadn't been consumed in the blast, knowing that such residue traces can remain at the scene for long periods.

Amid the scattered wreckage, crime scene investigators quickly found a piece of hot metal bearing a vehicle identification number that agents immediately traced to a missing Ryder rental truck. This single bit of evidence rapidly set in motion a series of further identifications. It quickly led agents to Elliott's Body Shop in Junction City, Kansas, who had served as the rental agency.

The person who signed the rental agreement had identified himself as BOB KLING, SSN: 962-42-9694, South Dakota's driver's license number YF942A6, and provided a home address of 428 Malt Drive, Redfield, South Dakota. The truck renter listed his destination as 428 Maple Drive, Omaha, Nebraska. FBI investigators quickly determined all of this information was bogus.

IDENTIFIERS OF THE DEAD

The leading identifiers of unidentified persons or partial human remains include dental records, physical description, tattoos, fingerprints, documentation found on the body, jewelry, and other personal effects. In recent years, identification by DNA is more prevalent. Without forensic evidence, many missing persons would never be positively identified.

Meanwhile, based on a description provided by an eyewitness at the rental agency, a forensic artist from the FBI's Investigative and Prosecutive Graphics Unit developed composite sketches known as "artists' impressions" of two suspicious individuals whom the witnesses said had been together renting the truck. These composite "John Doe" drawings of Unsub. #1 and Unsub. #2 were quickly copied and circulated around the world. They appeared on the front page of every major newspaper in the United States.

On April 20, 1995, FBI agents interviewed three witnesses who had been near the scene of the blast. They were shown a copy of the composite drawing of Unsub. #1 and identified him as closely resembling a person the witnesses had seen in front of the federal building before the explosion. The witnesses advised the FBI that they observed a person identified as Unsub. #1 at approximately 8:40 a.m. when they entered the building. They again observed Unsub. #1 at approximately 8:55 a.m., still in front of the 5th Road entrance of the building when they departed just minutes before the explosion.

Agents showed the composite drawings to employees at various motels and commercial establishments in the region. Employees of the Dreamland Motel in Junction City, Kansas, advised them that an individual resembling Unsub. #1 had been a guest from April 14 through April 18. This individual had registered under the name of Tim McVeigh, listed his automobile as bearing an Oklahoma license plate with an illegible plate number, and provided a Michigan address, on North Van Dyke Road in Decker, Michigan. The individual was last seen driving a car described as a Mercury from the 1970s.

A check of Michigan motor vehicle records showed a license in the name of Timothy J. McVeigh, date of birth April 23, 1968, with an address of 3616 North Nan Dyke Road, Decker, Michigan. This Michigan license was renewed by McVeigh on April 8, 1995. McVeigh had a prior license issued in the state of Kansas on March 21, 1990, and surrendered to Michigan in November 1993, with the following address: P.O. Box 2153, Fort Riley, Kansas.

Further investigation showed that the property at 3616 North Van Dyke Road, Decker, Michigan, was associated with James Douglas Nichols and his brother Terry Lynn Nichols. The property was a working farm. Terry Nichols had formerly resided in Marion, Kansas, approximately one hour from Junction City. A relative of James Nichols reported to the FBI that Tim McVeigh was a friend and associate of James Nichols, who had worked and resided at the farm on North Van Dyke Road in Decker. This relative further reported that she had heard that James Nichols had been involved in constructing bombs back around November 1994, and that he possessed large quantities of fuel oil and fertilizer.

Sketch of Unsub. #2, May 1995. Several witnesses reported seeing this man with McVeigh in the days before the bombing. Was it José Padilla, the "dirty bomb" conspiracy suspect?

On April 21, a former coworker of McVeigh's reported to the FBI that he had seen the composite drawing of Unsub. #1 on the television and recognized the drawing to be Tim McVeigh. He further advised that McVeigh was known to hold extreme right-wing views, was a military veteran, and was particularly agitated about the conduct of the federal government in Waco, Texas, in 1993.

In fact, the coworker further reported that McVeigh had been so agitated about the deaths of the Branch Davidians in Waco, on April 19, 1993, that he personally visited the site. Afterwards, McVeigh reportedly expressed extreme anger at the federal government and said that the government should never have done what it did. Investigators soon realized that the bombing had occurred on the two-year anniversary of the Waco disaster.

On April 21, 1994, investigators learned that a Timothy McVeigh was arrested at 10:30 a.m. on April 19, in Perry, Oklahoma, for not having a rear license tag and for possession of a concealed handgun. He had been stopped approximately 78 minutes after the explosion and 75 miles away from Oklahoma City. He had been driving a mustard-yellow 1977 Mercury Marquis. He was still in custody. At the time of his arrest McVeigh gave his home address as 3616 North Van Dyke Road, Decker, Michigan, and James Nichols of Decker, Michigan, as a reference.

McVeigh and Terry L. Nichols were arrested in connection with the bombing. McVeigh's car was found to have been loaded with anti-government propaganda, and according to the FBI his pockets, jeans, and earplugs revealed traces of PETN (Pentaerythritol Tetranitrate, also known as Penthrite), a compound used in detonator cord. The FBI laboratory quickly reported it had found ammonium nitrate crystals embedded in the truck. Analysts later estimated that the Ryder's load had contained 4,800 pounds of ammonium nitrate explosive—a massive fertilizer bomb.

The FBI never identified Unsub. #2, but Web sites soon pointed out his face bore a remarkable resemblance to that of José Padilla, whom the government had arrested as the "dirty bomber" allegedly involved with the September 11, 2001, terrorist attackers. Padilla, also known as Abdullah al-Muhajir, is currently held without charge in a naval brig at Hanahan in South Carolina as an "illegal enemy combatant," despite having been born in Brooklyn. Some subsequent reports said the FBI artist's sketch of Unsub. #2 was based on misinformation and never should have been released.

During McVeigh's trial, a whistleblower alleged gross deficiencies in the FBI Crime Laboratory. Some of the disclosures cast doubt on the FBI lab's early reports about finding ammonium nitrate traces and other key evidence. But McVeigh was convicted, sentenced to death, and executed by lethal injection in 2001. He was the first federal prisoner to be executed in 38 years. Nichols was convicted and sentenced to life imprisonment.

The Oklahoma City bombing turned out to be a classic identification case, but the quick conviction of McVeigh and Nichols left many unanswered questions and conspiracy theories. Some of these questions, involving the FBI Laboratory, are examined elsewhere in this book.

THE SCIENTISTS

Clyde Snow, Ph.D.—forensic anthropologist, "osteobiographer," and authority on human bones. Some of his famous skeletal studies have included John F. Kennedy, the men who fought in General Custer's "last stand" in 1876, Dr. Josef Mengele, the famous Nazi war criminal who fled to Brazil, the Argentinian "disappeared" (above), victims of serial killer John Wayne Gacy, the Egyptian boy king Tutankhamun, and victims of the Oklahoma City bombing.

Learning from the Dead

The rapid emergence of insect evidence has helped crack some cases but forensic entomology may still be in an early stage: chrysalid perhaps, or even larval. The autopsy is of course often the key part of many forensic investigations into unexplained death; and the pathologist is looking for a lot more than just bugs. There is a sign that is sometimes found where autopsies are performed. In Latin it is "Hic locus est ubi mors gaudet succurrere vitae," which translates as:

"This is the place where death rejoices to teach those who live."

This undated photograph from the Library of Congress archives is labeled, "Dynamite victim at Morgue Landing, Balt." What a challenge that would represent to a modern-day forensic pathologist: time of death, cause of death, identity . . . Immersion in water brings its own analytical challenges. In Azerbaijan in 1962 a decomposed body was found in a saltwater tank. Lab experiments on live fly larvae found on the body indicated that they could not survive for long in salt water, thus proving that the death had occurred elsewhere. PMI was estimated at seven to ten days. The killer's confession confirmed it was nine days—and 24 hours in the tank.

Bones: Forensic Anthropology

Bone detectives can tell a lot from skeletal remains.

When a human skeleton or apparent human bone is discovered outside a proper graveyard the police may call in a forensic anthropologist to identify the remains, help to determine an approximate postmortem interval, and determine if foul play may have been involved.

Whether the bones were discovered by a cadaver dog, a passing hunter, or a police search team, the investigator must be careful not to disturb the remains in any way.

The forensic anthropologist may work alone or with a trained evidence collection team. As with an archaeological dig, the first step is to mark off the area with a grid so that materials taken from each section can be collected, catalogued, and photographed for later reconstruction of the site. Surface materials such as leaves, plants, and sticks are removed and examined for any evidence.

The skeleton is carefully exposed using tools ranging from garden trowels to remove large amounts of dirt, to teaspoons and small brushes to remove the remaining soil and other debris. The bones are then removed and transported to the laboratory where the forensic anthropologist can examine them and listen to their story.

Forensic anthropologists apply standard scientific techniques developed in physical anthropology to identify human remains, and to assist in the detection of crime. They often work in conjunction with homicide investigators, forensic pathologists, and odontologists to suggest the age, sex, ancestry, stature, and unique features of a decedent, based on the bone evidence from the crime scene. Bone detectives may study everything from war crimes and battle scenes to apparent victims of murders, plane crashes, fires, and natural disasters.

By studying the pelvis, base of the skull, forehead, and jaw, a specialist can determine if the person was a male or female. Women usually show a wider pelvis and males a more prominent brow ridge, eye sockets, and jaw.

Examination of the joints, bones, and teeth can help to gauge the subject's age. The smoother the skull, the older the person. The spinal column also provides age-related clues. A child's skull has more separation between the bone plates. Knowledge of wrist development for children under thirteen can also help to determine some children's age.

Measurements of the length of leg and arm bones can help to determine the body's approximate height, and they carefully study the femur and metacarpals in the hand. By examining the wear on the bones at certain points, they may get a clearer sense of the person's likely weight. The width and height of the nose, along with facial or head hair, may help to identify the subject's race, although today's physical anthropologists tend to avoid using the term. By spotting where there is more muscle attachment to the bones on the dominant side, a physical anthropologist can tell whether the person was right- or left-handed. He or she can tell if the subject ever suffered a bone fracture, and this may later be compared to X-rays from missing

The Mütter Museum in Philadelphia contains an extraordinary collection of medical oddities but is nevertheless a repository of serious medical and physiological research. The forensic pathologist naturally relies on the knowledge gained over the centuries through autopsy to guide him or her in assessing possible victims of capital crime.

persons. The forensic anthropologist detects any signs of trauma, such as stab marks, cracks on the skull, crushed bones, and bullet holes. A fractured hyoid bone may indicate that the person was strangled.

If soft tissue is still present, this can help to determine the approximate time of death. Females lose one pound of tissue a day during decomposition; males three pounds a day. Acidic soil accelerates decomposition; alkaline soil retards it. A dead body in the early stages of decay can give off as many as 400 different chemical odors, so the anthropologist may wish to cover his or her nose.

Taken to the lab for closer analysis, the bones may reveal a lot more. When studied under a microscope, the skull fracture may end up being matched to the tire iron that served as the murder weapon, showing clearly observable patterns and irregularities, much like the rifling marks on a bullet. The stab wound may have left distinctive marks on the bones, telling whether it likely came from a machete or a stiletto. Multiple marks may reveal elements of the crime, or show that the victim suffered extensive defensive wounds from fighting off his or her attacker.

One noted forensic anthropologist, William Maples, once said, "The science of forensic anthropology consists of listening to the whispers of the dead."

Bugs: Alton Coleman and Debra Brown

In 1984 when police in Waukegan, Illinois discovered the bound and decomposing body of a young girl in an abandoned warehouse, they knew they had a strangling homicide, but their crime scene investigators were hampered by swarms of blow flies, maggots, and other insects that were feeding on the victim's bloated flesh.

Armed with a fingerprint taken from the crime scene, investigators soon came to suspect that the girl's sexual assault and murder was one of at least eight murders, seven rapes, and fourteen armed robberies that had been carried out in several Midwestern states by Alton Coleman, a severely troubled and sadistic individual, and his twisted girlfriend, Debra Brown. To help them salvage evidence from the corpse and the crime scene, the FBI contacted the Field Museum in Chicago, whose staff put them in touch with a leading entomologist at the University of Illinois at Chicago, Professor Bernard Greenberg.

Greenberg's original training was in acarology, the study of mites, but he had gone on to become a world authority on the more than 1,000 species of blow flies, or carrion flies, making up the family Calliphoridae. Based on his scientific research, he was acutely aware of the insects' biology and life cycles. So, as Greenberg studied detailed police photographs from the crime scene, he discovered a festering of at least three species at various stages of development, and he commenced an intensive project that would try to utilize bugs, not as a hindrance to the investigation, but as a crucial source of forensic evidence.

Homicide investigators were already aware that flies are immediately attracted to a dead body. Greenberg explained he knew the creatures can detect carrion within minutes and then lay eggs on the exposed flesh within a couple of hours. After these eggs are laid, the young insects pass through a series of larval and pupal stages that are so remarkably predictable that a skilled scientist can use this information to calculate the insects' age and hence deduce the approximate time of death. Such an approximation is dependent on time of day, time of year, temperature, corpse exposure to soil or water, and other factors. However, by analyzing the time it took for the flies to reach specific stages of development, Greenberg said he was able to determine when the victim died within a two-day range. This forensic evidence, combined with eyewitness reports of a man seen walking with the girl just before her death, helped to gain Coleman's capital conviction. He was subsequently executed in Ohio.

Greenberg went on to become regarded as the "father of forensic entomology," spawning a tiny but rapidly emerging new specialty. Beginning in the late 1970s a handful of bug experts in Europe and the United States had begun to speculate about the possible application of insect evidence to criminal investigations, but Greenberg had quickly developed such knowledge with such assurance that the courts allowed his evidence to help justify the ultimate punishment.

In custody Debra Brown discussed the bloody crime spree in detail. In court, the defense argued that her Fifth Amendment right against self-incrimination had been violated, which was in part upheld. But she also was condemned to death.

In fact, a form of forensic entomology had been practiced in China as early as the thirteenth century, when a Chinese death investigator, Sung Ts'u, wrote a book entitled *The Washing Away of Wrongs*, which was translated into English by B. E. McKnight in 1981. Sung told the ancient story of a murder in which the victim was repeatedly slashed and the local magistrate concluded the wounds had been inflicted by a sickle. In an effort to solve the crime, the magistrate ordered all the village men to assemble, each with his own sickle. In the hot summer sun, flies were soon attracted to one sickle, due to the invisible residue of blood and carrion clinging to the blade. The murderer soon confessed. Sung's writing also discussed the relationship between maggots and adult flies that invaded the dead.

Greenberg and other entomologists began to apply modern knowledge about arthropods—invertebrates with a segmented body and paired, jointed legs, such as the insects, the mites, and the spiders—that are present on a decomposing body, to help determine the postmortem interval (PMI) and the site of the death. He and other scientists began to establish that insects are attracted by specific states of decay and that particular species colonize a corpse for a limited period of time. This produces a faunal succession on cadavers. Close analysis of the bug evidence, together with the knowledge of their growth rates under specific environmental conditions, helps the forensic entomologist estimate time since death.

PUPAL PUPILS

By 1996 a group of scientists had established their own American Board of Forensic Entomology in the United States and Canada. They pointed out that insects can provide a useful forensic tool in a homicide case in other ways besides helping to establish PMI. For example, forensic entomology may help to reveal the following:

- If the body has been moved, and if so, by determining the insects' native habitat, it can indicate the type of area where the murder actually took place.
- If the body has been disturbed.
- The presence and position of wounds that otherwise may have become undetectable due to decomposition.
- The presence of drugs in the body.

Research has indicated that insect evidence can continue to generate at a crime scene for an extended period after the corpse was removed and even after the apartment or other area was extensively cleaned. Insects from the crime scene may also later be found on someone who has visited the body. Entomological evidence has also been used in product tampering, food contamination, child abuse cases, and sexual assaults.

Temperature and other environmental factors are extremely important to forensic entomologists and they must obtain extensive data about these variables to support their conclusions. Forensic entomologists must seek to find out how much sun the site has received, what time of day the body may have been placed there, what scavengers inhabit the area, and many other facts that can affect their conclusions. If the body was frozen or

THE BODY FARM

In 1972 Dr. William M. Bass founded the Anthropological Research Facility (ARF), otherwise known as the "Body Farm," at the University of Tennessee at Knoxville, to provide donated human remains and animal carcasses for forensic anthropologists to use in their study of the process of decomposition. As many as 40 or so rotting bodies in various stages of putrefaction are scattered throughout the facility's grounds, some of them stuffed into car trunks or buried under layers of leaves. Researchers study the rate of decay, the infestation by insects, and other morbid aspects that can help them in their forensic work. The site is also utilized by the FBI and mystery writers.

buried deeply, most insects may have been excluded, although even a body or body part that was found securely sealed in a garbage bag may turn out to be maggot-infested and full of insects.

The multiplicity of factors affecting the presence and rate of insect infestation make conclusions about entomological forensic evidence very subjective. It is not like DNA. This subjectivity increasingly has led to "showdowns" between expert witnesses called by the prosecution and the defense. In 2002, for example, the defense attempted to utilize insect evidence in behalf of David Westerfield who was on trial in San Diego for the brutal sexual assault and murder of seven-year-old Danielle van Dam. Her body was found four weeks after her disappearance in a dry, brushy area of East County on February 27. Westerfield was ultimately convicted and sentenced to death. Bug evidence couldn't save him. It was more conventional forensic evidence that convicted him: fingerprints on a cabinet, blood traces on a jacket, and a single orange fiber. "This was a forensics case, ladies and gentlemen, and there was heroism from the men and women at the lab who put this case together," said District Attorney Paul Pfingst.

Autopsy

The skilled forensic pathologist tries to give the crime victim a voice in solving the crime. Sometimes the best source of information about how a person died is the lifeless body itself.

The question is, how to unlock such secrets? This is the function of the forensic autopsy. The term derives from the Greek meaning to "see for oneself" or to "see with one's own eyes." The dissection of cadavers has been going on for centuries, usually for scientific or educational, medical purposes. If properly done, today's autopsy can take some of the most important evidence from the crime scene to draw startingly detailed conclusions about the cause of death and the precise nature of the crime. A top forensic pathologist is worth his or her weight in gold.

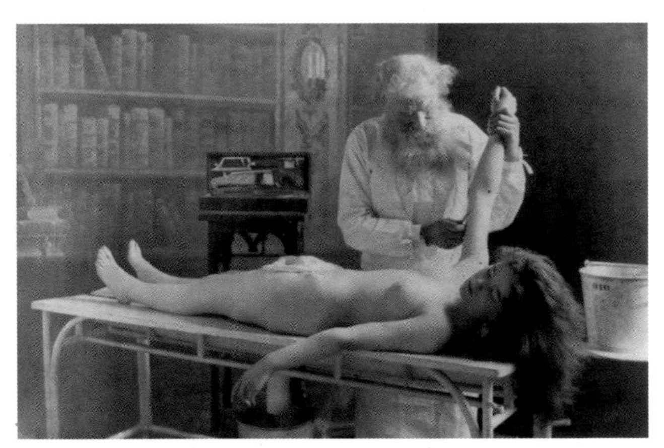

An unconvincing photograph of an autopsy, taken in 1900 by Fritz W. Guerin. Guerin sometimes posed models for medical scenes. The bucket—even the pathologist—may be for real. The "corpse" may be playing possum.

Although often reviled throughout history as the ultimate form of exploitation or assault, autopsies have recently taken on an aura of infallibility, as the fount of all criminalist knowledge, spawned by best-selling books about medical detectives and popular television shows such as *C.S.I.*

Contrary to public perception, however, the number of autopsies performed in the United States has actually been steadily decreasing since 1955. In the 1940s, as many as 50 percent of all deaths resulted in an autopsy, but by 1985 the figure was only 14 percent. In 1995, the National Center for Health Statistics stopped collecting autopsy statistics altogether. Part of the reason for the decline is attributable to the advent of other less invasive medical procedures such as magnetic resonance imaging (MRI). But when it comes to homicides or other potentially unnatural deaths, autopsies are legally required. And despite the drop in overall use, the persons ruling on cause of death and conducting autopsies are also better trained and equipped than in previous eras.

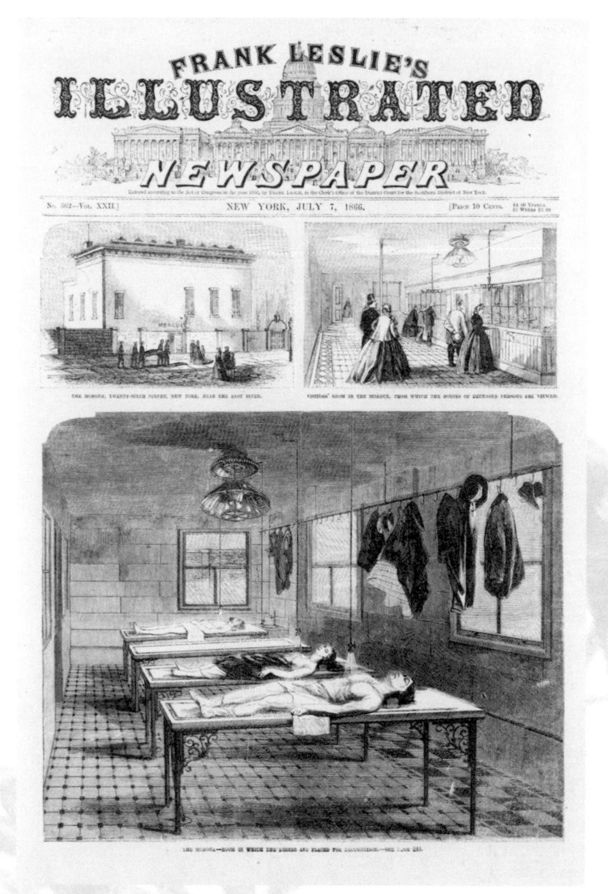
FRANK LESLIE'S ILLUSTRATED NEWSPAPER

No. 562—Vol. XXII.] NEW YORK, JULY 7, 1866. [Price 10 Cents.

The exterior of the New York morgue on 26th Street and interior viewing rooms, 1866. Bodies found in the East River often remained unidentified. Brooklyn only began registering births in that year.

The old coroner system that operated throughout the nation didn't require coroners to have any medical training, so that very few of them were physicians. In fact, most coroners were undertakers or politicians. But today, at least half of the jurisdictions rely on a medical examiner (who is a medical doctor) to fix legally the cause of death.

There are only a few hundred forensic pathologists—physicians who are specially trained to be able to determine a cause of death and later support their conclusions in court—assisting the coroners or medical examiners. (Forensic pathologists take an examination given by the American Board of Pathology.)

In the case of a murder or suspicious death, the body arrives at the medical examiner's office or hospital morgue in a body bag or evidence sheet that is designed to prevent contamination. Led by a designated chief, a team of specialists with protective clothing and equipment works together to conduct the procedure, with each person assigned to carry out specific tasks.

When the body is removed from its container and put on a metal table, care must be taken to remove any forensic trace evidence that accompanied it from the crime scene. In the case of the

J.F.K. assassination, for instance, a stretcher used to transport Governor Connally was later found to carry the bullet that would be blamed for wounding him and killing the President.

Once the cadaver is on the table, it is photographed as it looked when it arrived. Notes are taken and sometimes videotape is shot to record every observation. Some autopsy rooms are equipped with hanging microphones connected to voice recorders that can be activated by a foot pedal. The subject's clothes are taken off and examined, and they too are photographed and stored. Any undressed wounds are closely examined. The body is then fingerprinted and the face is photographed to establish positively the victim's identity.

Then evidence such as gun powder residue, paint flakes or carpet fibers are collected from the external surfaces of the body. Members of the autopsy team take fingernail scrapings, blood and hair samples, and vaginal and rectal swabs, all for testing purposes. Ultraviolet radiation or a laser light may be used to search the body for evidence that may not be easily visible to the naked eye.

The body is then washed and photographed again and X-rays may be taken. After the body is cleaned, it is weighed and measured before being placed on the autopsy table. Data about the subject's age, race, sex, hair color, eye color, and other identifying features are recorded into a hand-held voice recorder or notebook to aid in the preparation of standard forms and autopsy reports.

Members of the autopsy team thoroughly examine the body from head to toe, scouring it for bruises, wounds, needle marks, bullet holes, ligature marks, and other signs of injury that may not have been noticed at the crime scene. Some marks might be classified as bite marks or simply as lividity.

The internal examination begins with a large and deep Y-shaped incision that is made from shoulder to shoulder meeting at the breastbone and continuing all the way down the torso to the pubic bone. The skin, muscles, and soft tissue are then peeled back to expose the inner organs, with the upper chest portion pulled up over the face to expose the ribcage. Further incisions are made on each side of the ribcage to pull the ribcage from the rest of the skeleton.

THE SCIENTISTS

Thomas T. Noguchi, M.D.—former Chief Medical Examiner-County Coroner of Los Angeles County, 1961 to 1982, he became known as the "coroner to the stars" after performing autopsies on Marilyn Monroe, Robert Kennedy, John Belushi, Sharon Tate, Janis Joplin, and other celebrities; author of the best-selling book, *Coroner* (1983). Here he takes notes while examining one of the two bodies found on the lawn of the Tate-Polanski home, Bel Air (1969).

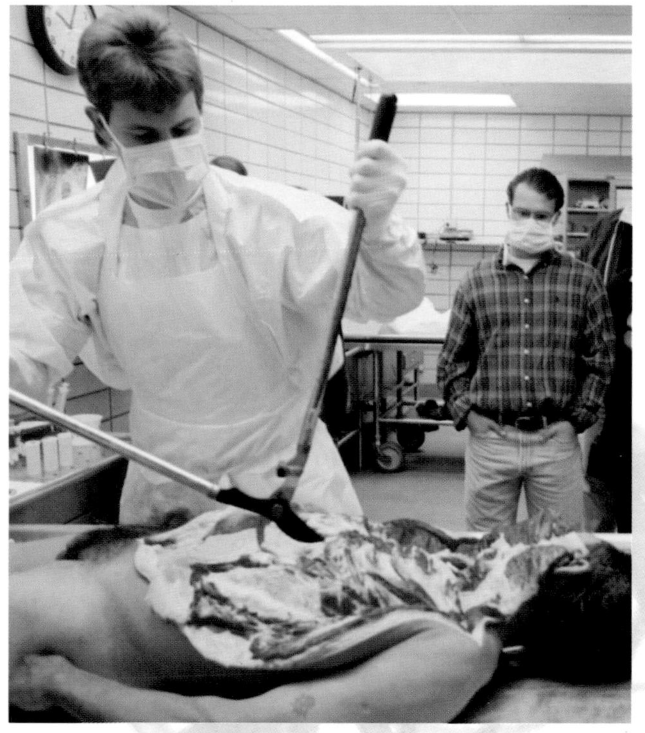
The Deputy Medical Examiner performs an autopsy before a Law School class, Houston, Texas. The homicide victim was shot to death that morning. The ribcage is being incised to expose the internal organs.

Once all of the internal organs are exposed, a series of cuts are made to pull out major blood vessels so they can be inspected and noted. The stomach is removed and its contents are weighed and examined to help determine what the subject had consumed, and when. This could be especially useful. Along with the extent of algor mortis (cooling of the body), rigor mortis, or livor mortis (the settling of the blood in the lower portion of the body, which causes a purplish discoloration of the skin), and stage of decomposition, this can help the forensic scientist in the estimation of the postmortem interval (PMI), or the time that has elapsed since death occurred.

The other organs, including the brain, are also examined and sampled. Standard toxicological tests must be performed to determine the presence of alcohol, cocaine, barbiturates, or other drugs or combinations of drugs. Important specimens must be properly labeled and preserved in jars.

After the body has been sewn back up, the medical examiner must prepare the documentation, determining the relative health of the person before they died, along with any appropriate diagnosis, and facts supporting the official cause of death.

Completion of an ordinary, uncomplicated autopsy for a person who has died of natural causes may take about two hours to perform, whereas an especially complicated and legally sensitive postmortem examination may take as long as six or eight hours. Compared to what could be discovered and supported in the 1930s or even the 1960s, today's forensic autopsy benefits from many enormous advances in the medical field. Even basic instruments have improved. While the rib cutters shown above right have not changed, the vibrating, or "Stryker" saw allows for much more accurate and smaller incisions. The very fast back-and-forth movement of the blade helpes to avoid damge to soft tissues. A typical autopsy report may run to five or so pages in length, stating facts that support the cause of death and manner of death.

Autopsy/Ballistics: Robert F. Kennedy

The handgun assassination of Senator Robert F. Kennedy in a crowded Los Angeles hotel pantry triggered a dispute between the medical examiner who performed the autopsy and the LAPD's chief criminalist that simmers to this day. When Presidential candidate Robert F. "Bobby" Kennedy was fatally shot and five bystanders were wounded at the Ambassador Hotel on June 4, 1968, the act was attributed to a lone assassin, Sirhan Sirhan, a Palestinian Arab, who was forcibly disarmed at the scene after he had emptied his eight-shot revolver at RFK.

Under the law at the time, the crime fell under the jurisdiction of Los Angeles authorities, not the federal government or inexperienced military doctors. From 3 a.m. to 9:15 a.m. on the morning of July 6, Los Angeles Coroner and Chief Medical Examiner Thomas Noguchi and two of his assistants, Dr. John E. Holloway and Dr. Abraham T. Lu, conducted the official autopsy on the body of the slain victim. The dozen others who witnessed the procedure included the physician who had tried to save Senator Kennedy's life, three top pathologists from the Armed Forces Institute of Pathology in Washington, the three chiefs of the Institute's Ballistic Wound branch, Forensic branch and Neuro-Pathology branch, as well as representatives of the FBI, Secret Service, LAPD, and Sheriff's office.

Noguchi concluded that the bullet lodged in the senator's brain stem had entered from behind the right ear at very close range. If RFK did not turn further than in profile to the shooter, there is a major discrepancy.

Noguchi, considered one of the nation's most experienced and savvy medical examiners, removed one intact bullet and fragments from another. In his 62-page report, he later stated that the shot that killed RFK "had entered through the mastoid bone, an inch behind the right ear and had traveled upward to sever the branches of the superior cerebral artery." The largest fragment of that bullet lodged in the brain stem. Another shot had penetrated Kennedy's right armpit and exited through the upper portion of his chest at a 59-degree angle. A third shot entered one-and-a-half inches below the previous one and stopped in the neck near the sixth cervical. He said this was the bullet that was found intact.

After also examining Kennedy's clothing, Noguchi announced that a fourth bullet had been fired at the senator. He said it entered and exited the fabric without touching the senator.

Faced with mounting pressure to explain how eight bullets fired from one gun could account for all of the damage, the LAPD reported that the weapon wrested from Sirhan was an Iver-Johnson .22-caliber handgun, the chamber of which had expended all eight cylinders. On July 8, the LAPD's chief criminalist of the Scientific Research Division, DeWayne A. Wolfer, released a bullet-accountability report that stated the following:

RFK discusses school with a future voter, Brooklyn, 1966. For conspiracy theorists, the question must be, if not Sirhan Sirhan alone, then who else? *Cui bono?* Who stood to gain from stopping the future president?

- Bullet #1: struck Senator Kennedy behind the right ear and was later recovered from his head.
- Bullet #2: passed through RFK's right shoulder pad and struck campaign aide Paul Schrade in the forehead, from which it was later recovered.
- Bullet #3: entered RFK's right rear shoulder and was later recovered from his 6th cervical vertebrae.
- Bullet #4: entered RFK's right rear back, about one inch below bullet #3, but exited the senator's body through the right front chest and was lost somewhere in the ceiling interspace.
- Bullet #5: struck victim Ira Goldstein in the right rear buttock from which it was recovered.
- Bullet #6: passed through Goldstein's pants leg, struck the cement floor, and ricocheted into Irwin Stroll's left leg, from which it was recovered.
- Bullet #7: struck William Weisel in the left abdomen, where it was recovered.
- Bullet #8: reflected off the plaster ceiling to strike victim Evans in the head from which it was recovered.

The LAPD also concluded that a Walker's H-acid test was conducted on Senator Kennedy's suit coat and it determined that the muzzle of the weapon was held at a distance of one to six inches from the coat at the time of all the firings.

But Noguchi responded that in his opinion there was really no way to accurately trace the flight pattern of so many bullets. He also concluded that the shot that killed Senator Kennedy was fired into the back of his neck at a range of no more than one-and-a-half inches, for it left thick powder burns on his skin. Yet all of the witnesses claimed that Sirhan Sirhan shot his weapon in front of Kennedy, giving more support to the scenario of a second shooter firing from the rear. A decade after the murder, Noguchi added more fuel to this controversy when he wrote, "Until more is precisely known … the existence of a second gunman remains a possibility. Thus, I have never said that Sirhan Sirhan killed Robert Kennedy."

Some conspiracy theorists also pointed out that, in addition to the eight bullets accounted for in the LAPD report, both LAPD and FBI personnel identified two additional bullets as lodged in the door frame of the pantry. But the door frames were later burned and the LAPD said it had also destroyed the records of its tests on the "bullet holes" in the doorframe.

Was it a cover-up? Sirhan remains in prison for the murder. But based on some of the forensic evidence, startling questions continue to haunt the case.

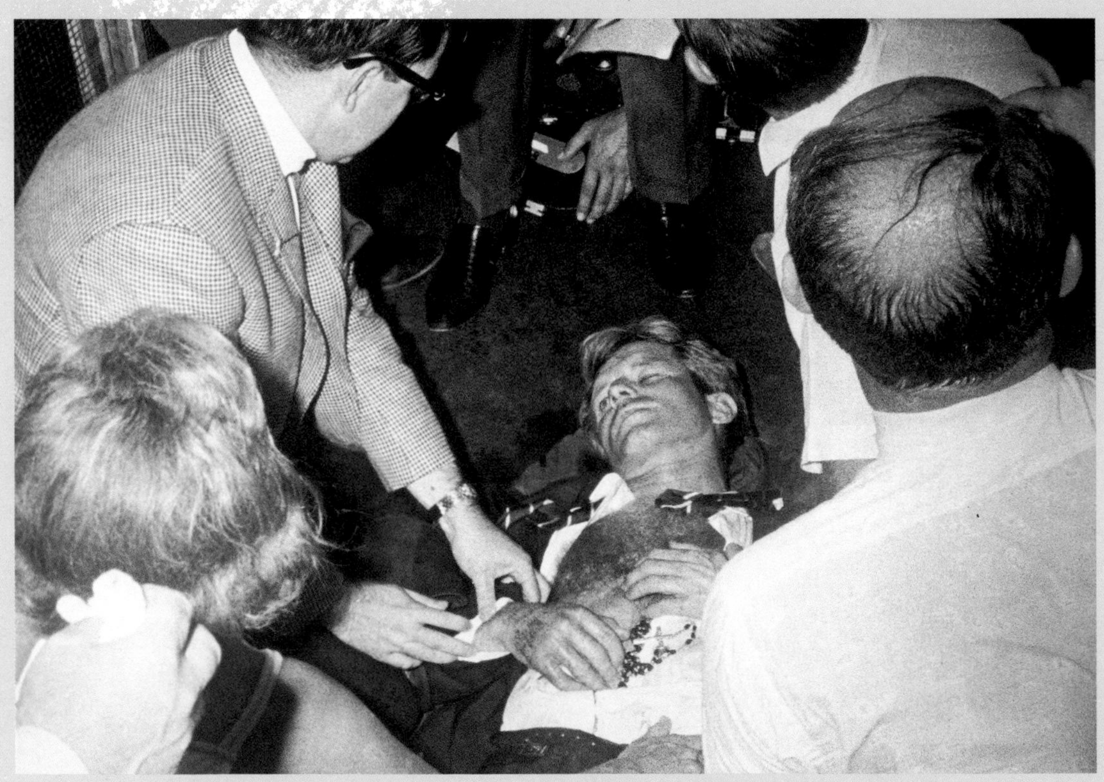

Robert Kennedy lies on the floor of the Ambassador Hotel, clutching his rosary beads. His wife Ethel is bottom left. Sirhan's apparent motive is not hard to find: the Palestinian felt personally betrayed by Kennedy's support for Israel in the June 1967 Six-Day War. Did he act alone, or was there a conspiracy to kill the President-to-be? Was Sirhan "programmed to kill"?

Facial Reconstruction

Facial reconstruction involves a combination of science and art to identify someone's face based upon skeletal remains.

In the 1930s, the FBI began to utilize Ales Hrdlika (1869–1943), a renowned physical anthropologist at the Smithsonian Institute, to try to help identify apparent crime victims based on skeletal specimens. Hrdlika didn't publish anything about his findings, but on at least 37 occasions, he offered opinions to the Bureau about whether remains he examined were of human origin, and if so, he commented on the subject's apparent age at death, antiquity, sex, stature, ancestry, and evidence of foul play. Over the years, Hrdlika's successors at the Smithsonian continued to collaborate with the FBI, advising law enforcement on a growing number of cases.

Some bone specialists tried to take their work a step farther by carrying out a facial reconstruction of the victim's face. At least one clay facial reproduction had led to the identification of a victim in a New York case as early as 1916. But America's top physical anthropologists remained skeptical of the practice. In 1978 Betty Pat Gatliff, a medical illustrator and sculptor, was hired to produce facial reconstructions of nine unidentified victims of John Wayne Gacy, the Illinois serial killer, but the release of the images brought no identifications.

By the early 1980s, a few leading forensic anthropologists began to express more confidence in their ability to undertake facial reproduction and some began to use computer imaging techniques. Photographic superimposition techniques also began to be used in missing person cases where a recovered cranium was thought to belong to a specific missing person for whom photographs were available. Efforts by the FBI to undertake such reconstructions peaked in the early 1990s but ended in 1996, as new molecular approaches became more effective.

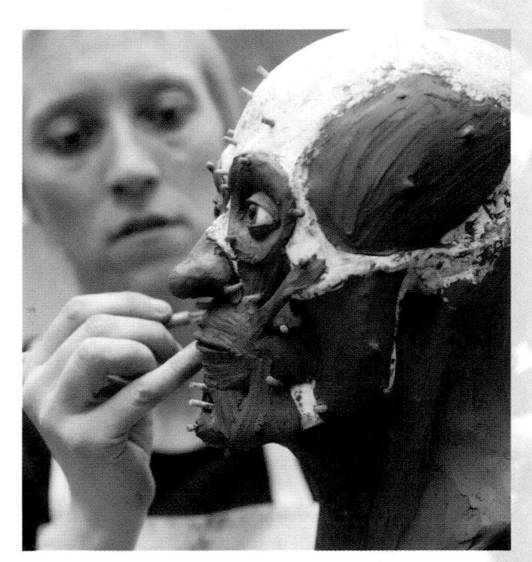

Pegs representing the depth of facial muscles are attached to the cast of an unidentified skull to guide the artist.

Forensic anthropologists were also called in to consult in helping to reconstruct the appearance of long-lost fugitives. In 1989, for example, forensic anthropologist Frank A. Bender of Philadelphia prepared a computer-generated likeness of a murder suspect whom law enforcement authorities had not seen for 18 years. Bender's three-dimensional image later led to the suspect's capture and he was ultimately convicted.

Ballistics

Ballistics is the study of the dynamics of projectiles; and for forensics, that means firearms—functioning, firing, flight of ammunition, and effect. The study of the effect can be subdivided as terminal ballistics, which is just the right term for what often faces the forensic pathologist.

Women familiarizing themselves with police firearms issue in 1915, at the time of the infamous Stielow case in Shelby, NY. The farmer was convicted of a double murder and sentenced to death on the evidence of a firearms "expert" who linked scratches on the bullets fired to Charles Stielow's revolver. NYC lawyer and suffragist Inez Millholland campaigned on his behalf. Charles E. Waite, a special investigator for the New York Attorney General's Office, exposed the bogus ballistics on appeal and helped save Stielow's life.

Ballistics: President John F. Kennedy

The JFK assassination lives on as history's most famous ballistics case—full of blunders leading to improbable, but maybe correct, conclusions.

The assassination of President John F. Kennedy in Dallas on Friday, November 22, 1963, and its aftermath stunned the nation as much as any event in American history. The case stands as one of the significant watersheds in forensics—ranked at or near the top of the list for its cutting-edge contributions in ballistics, forensic pathology, photography, and acoustics, to name a few of the areas most affected. Although the assassination has been investigated more intensively than any other crime, most Americans continue to disbelieve the Warren Commission's key finding that there was no conspiracy to kill the President.

Millions of television viewers saw reports and images of the shooting of the President and Texas Governor John B. Connally as they rode in an open-top limousine through Dealey Plaza

Who killed JFK? Surely no other murder or assassination has generated as much skepticism about the official verdict as his. Certainly, one of the reasons was the sloppy control of the scene of crime at the time and subsequent evidence collation. At a small price, anyone can examine 11,000 pages of Dallas PD files relating to the case on the Web. "Today at the police station they showed me a rifle. This was like the rifle my husband had. It was a dark gun." (Affidavit, Mrs. Marina Oswald, November 22, 1963.)

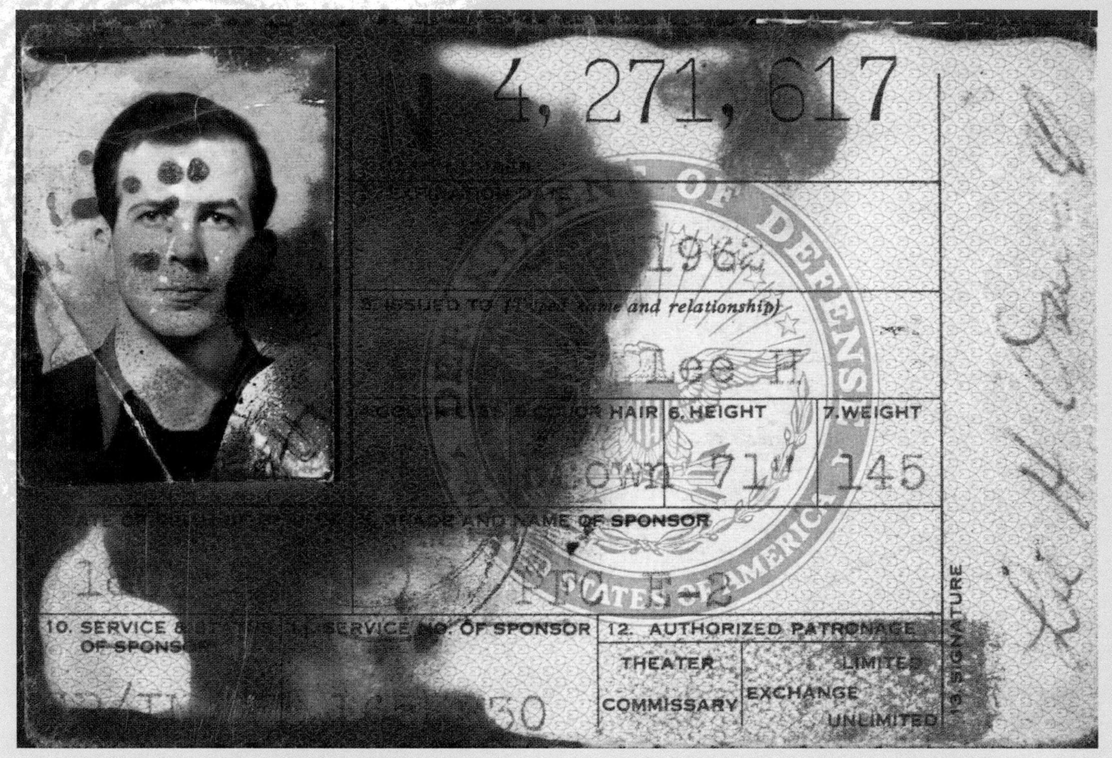

4,271,617

and relationship)

Lee H

HAIR 6. HEIGHT 7. WEIGHT

71" 145

OF SPONSOR

10. SERVICE & OF SPONSOR

11. SERVICE NO. OF SPONSOR

12. AUTHORIZED PATRONAGE

THEATER LIMITED

COMMISSARY EXCHANGE UNLIMITED

13. SIGNATURE

Not blood but FBI fingerprint fluid stains Oswald's military ID card, found in his wallet on the day of his arrest.

at 12:30 p.m. Kennedy would die shortly afterward and Connally was critically wounded but he would survive. (A spectator, James Tague, who had been standing 270 feet in front of Kennedy, suffered a minor gunshot wound to his right cheek.)

An eyewitness, Howard R. Brennan, located 107 feet from the depository at the time of the shooting, told police he saw a youngish man on the sixth floor with a rifle. He said he saw the rifle retreat from view after the shots were fired. Brennan gave a description to police—a white male, around thirty, slender build, height five feet ten inches, weight about 165 pounds—who sent it out over police radios at 12:45 p.m.

Police quickly discovered three empty shells found in a sixth-floor sniper's nest in the book depository, and a rifle later identified as the one used in the shooting was found hidden nearby. It was a 6.5 x 52mm Mannlicher-Carcano M91/38 bolt-action rifle, serial number C2766, with a side-mounted Ordnance Optics 4 x 18 scope. (Police would later determine it had been ordered through the mail under an alias, and they retrieved a palm print from the weapon.)

Shortly after the suspect's description was broadcast, at 1:50 p.m. the alleged assassin, Lee Harvey Oswald, was cornered in a movie theater, disarmed after a scuffle, arrested and taken to police headquarters. He was later paraded before the news media. Investigators quickly determined that Oswald had worked at the depository and had last been seen on the sixth floor carrying a long package he said contained "curtain rods."

Dallas police said Oswald had used a hand gun to murder a police officer, J.D. Tippit, who'd been trying to apprehend him at 1:15 p.m. Oswald was searched, put through a lineup, photographed, fingerprinted, palm printed, paraffin tested, and interrogated. (A paraffin, or dermal nitrate test, is a presumptive test for the presence of gunshot residue that was used for many years but is no longer used in forensic science.)

Later that night, based on growing circumstantial evidence but without a confession, Oswald was charged with murdering President Kennedy as well as Officer Tippit. The next day he was still trying to gain access to a defense lawyer. (This was before *Miranda v. Arizona.*) As things turned out, he would never get one.

Two days later, a man later identified as Jack Ruby shot Oswald to death in the police garage, in full view of television and still cameras. At the time, Oswald was handcuffed to Homicide Detective James R. ("Jim") Leavelle and images of the shooting became some of the most memorable news pictures in media history. It was also the first time that a homicide was broadcast live and at first American public opinion overwhelmingly supported Ruby's deed. The weapon Ruby used was a Colt Cobra .38 Special, snub-nosed revolver, serial number 2744 LW.

Ruby later claimed he shot Oswald on the spur of the moment when the opportunity presented itself, claiming his action would show the world that "Jews have guts," that he was helping the city of Dallas "redeem" itself in the eyes of the public, and that Oswald's death would spare Jacqueline Kennedy the ordeal of having to appear at Oswald's court trial. Ruby's murder conviction was later overturned but he died in 1967 while awaiting a new trial—after some conspiracy theorists had begun to question his apparent ties to organized crime and other oddities that had been overlooked by the government's investigations.

Today, all three crime scenes related to the JFK assassination would remain sealed off for an extended period to allow teams of forensic investigators to comb every inch for any scintilla of trace evidence. But in 1962 the grounds where the President had been shot were so poorly checked that the next day, a souvenir hunter walking his dog picked up a fragment of Kennedy's skull from the grass in Dealey Plaza. (He eventually turned it over to hospital officials but they somehow lost it.)

Part of the problem with the investigation stemmed from confusion over who was in charge. An investigation by Dallas police and other local authorities was essentially swept aside by federal authorities because the President had been shot. After arresting Oswald and collecting physical evidence at the crime scenes, at 10:30 p.m. on November 22, the feds told Dallas Police Chief Jesse Curry to send all of the physical evidence found, but not Oswald, to FBI headquarters in Washington. Then Ruby shot Oswald at Dallas police headquarters.

The FBI took just 17 days to complete its official investigation. But politics and history required a fuller accounting. On November 29, 1963, President Lyndon B. Johnson appointed a special panel, the Warren Commission, to probe the assassination. In late September 1964, after a 10-month investigation, the panel released its public report, concluding that it could not find any persuasive evidence of a domestic or foreign conspiracy involving any other person, group, or country, and that both Lee Harvey Oswald and Jack Ruby had acted alone.

The position that Oswald acted alone came to be known as the "Lone Gunman Theory." The Warren Commission also concluded that all three bullets fired during the assassination were fired by Oswald, from behind the motorcade in the Texas Book Depository. Its major findings included the following:

One shot (it could not determine which one) likely missed the motorcade.

The first shot to hit anyone struck Kennedy in the upper back, exited near the front of his neck and likely continued on to cause all of Governor Connally's numerous injuries (thus becoming known as the "Single Bullet Theory").

The last shot struck Kennedy in the head, destroying his brain and fatally wounding him.

Although the Warren Commission completed its work in 1964, many of its supporting documents were ordered held under wraps until 2017, thus contributing to public distrust about the truth and completeness of its findings.

A later House of Representatives Select Committee revisited the episode and found numerous deficiencies, but some of its findings were also criticized. Fourteen years after the assassination, Congress established the House Select Committee on Assassinations to attempt to get to the bottom of the controversy.

REASSESSING THE AUTOPSY

The House Select Committee hired Dr. Michael M. Baden as chief forensic pathologist and he assembled a panel of eight other well-respected medical examiners. Together they had performed over 100,000 medico-legal autopsies, many of them involving gunshot wounds. One of those Baden selected was Dr. Cyril Wecht of Pittsburgh, considered the leading forensic critic of the original autopsy.

One of the strongest criticisms of the federal government's investigation and the Warren Commission report centered around the autopsy. It had been performed by Naval hospital pathologists who had little, if any, experience of gunshot wounds—something that presumably would never be allowed to happen today. The chief doctor in charge of the autopsy, Commander James J. Humes, had no training in autopsies and he did not even bring in an experienced forensic pathologist—thereby bypassing many top specialists, including Earl Rose, the Dallas coroner, who legally had authority over the body.

After the autopsy was complete, Earl Rose had tried to prevent agents from removing the cadaver, so he could do his job, but Secret Service agents brandished weapons and used physical force to remove the body, fearing that Texas officials might be part of a conspiracy to overthrow the government.

Chief Justice Earl Warren of the US Supreme Court (left) swearing in John A. McCone as head of the CIA in April 1961. A poignant image: two years later, Warren would (reluctantly) head the investigation into the assassination of the onlooker, President Kennedy.

Incredibly, neither the FBI nor the Warren Commission had enlisted a competent forensic pathologist to assist them in their "exhaustive investigation" of the case. But Baden's star-studded team also had its problems. The House forensic team never got access to Kennedy's body—it was never exhumed. Instead, they had to rely on autopsy records, medical reports, photographs, X-rays, the President's and Governor Connally's bloodstained clothing, and other items.

They also had to cover every base, to address every kind of issue that had been raised or might be raised by conspiracy theorists. To their dismay they came to realize that many key pieces of evidence were missing—Kennedy's brain, pieces of his skull, microscopic slides of the tissues, some of the autopsy photographs, the paraffin blocks, and other items. Governor Connally's suit had been dry-cleaned and his Stetson hat and gold cufflinks were missing.

By interviewing some of the participants in the original autopsy, the House forensic team discovered that the abbreviated procedure had been conducted in an atmosphere of complete chaos, fear, and stress. FBI agents had thrown out of the autopsy room the hospital's trained photographer and substituted one of their own agent-photographers, who had no experience whatsoever with autopsies and not very much skill at photography. To their horror, the House team also found that the Warren Commission had never even examined the JFK autopsy photographs and X-rays.

Step-by-step, the House Assassination Committee team reviewed everything that had been done by medical personnel before the autopsy—information that had never been obtained by the physicians performing the autopsy. The original autopsy report was found riddled with serious omissions and errors of all sorts.

Nevertheless, the House team concluded that the original botched autopsy had reached the right conclusions—President Kennedy had been shot by two bullets from behind, one of which struck him in the back of the head and another hit his back.

The panel unanimously concluded from the X-rays that the fatal bullet had entered the rear of the President's head near the cowlick area and exited from the right front. No medical evidence was found to indicate that this massive wound was caused by a bullet fired from in front of, or from the side of, the President's car.

Despite exhaustive frame-by-frame analysis of the 8-millimeter color film of the assassination made by spectator Abraham Zapruder, which showed President Kennedy's head at the time of impact moving backward, not forward as might be expected, the panel concluded that neurological reactions could have caused the head to snap in any direction after being shot. Dr. Baden also pointed out that it was scientifically wrong to assert that the president's body should have jerked backward rather than forward, not only due to the force of the bullet, but also the sudden acceleration of the car after the driver had realized they were under attack.

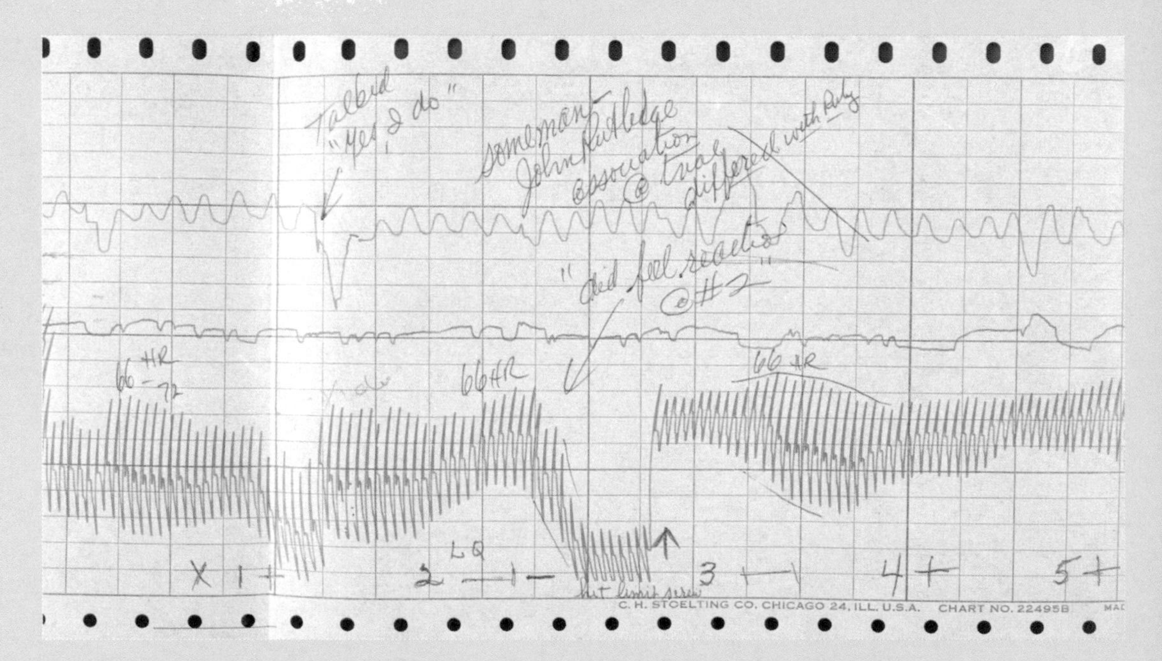

Jack Ruby's 1964 polygraph. The Warren Commission accepted the conclusion of FBI polygraph expert Bell P. Herndon that Ruby was truthful when he claimed not to know Oswald and to have had nothing to do with Kennedy's assassination (or more accurately, they took no notice either way). Others have argued unconvincingly since that the test was badly executed, even deliberately sabotaged. The testimony of Ruby's psychiatrist, Dr. William Beavers, who was present during the examination, was taken immediately after the polygraph examination. Having examined Ruby on about ten occasions, he diagnosed him as a "psychotic depressive."

ACOUSTICS

Besides subjecting the assassination to more up-to-date medical examination, the House Committee's investigation utilized new technology to analyze the available acoustical evidence, based on a recording that had been made at the scene. For several minutes before, during, and after the assassination a Dallas police motorcycleman's radio microphone was stuck in the "transmit" position and was recorded back at the police radio dispatcher's room on a Dictabelt.

To Dr. Baden's dismay, this acoustical evidence was used to override his medical team's findings. Counsel Robert Blakey used his acoustical analysis to support the Committee's new theory of a *likely conspiracy*. It concluded that Oswald fired three shots from behind Kennedy, hitting him with two of them and missing with one. Unlike the Warren Commission, it also found that an *unknown assailant* had fired one shot from the grassy knoll and also missed. The Committee decided that there was a probability of over 95 percent that there was a second shooter on the grassy knoll at Dealy Plaza.

After the FBI disputed the validity of the acoustic evidence the Justice Department enlisted the National Academy of Sciences to review it. In 1982 a panel of scientists, headed by Dr. Norman Ramsey, issued a report, which concluded that there was no compelling evidence for gunshots on the recording and that the House Committee's suspect signals had actually been recorded about a minute after the shooting happened.

There was another police radio channel recorded on the Dictabelt in the background. Interestingly, a critical role was played by S. Barber, a musician, who detected cross-talk between the police radio channels that proved the recording was made after the shooting.

BALLISTICS

The best evidence for identifying the alleged assassination weapon were the two bullet fragments found in the President's car and the nearly whole bullet found in a stretcher in Parkland Hospital in Dallas.

In 1964, FBI experts ballistically matched this bullet and fragments to the rifle barrel of the Mannlicher-Carcano by microscopic comparison of the markings in the barrel with those found on the bullet and fragments. A firearms panel of independent experts appointed by the House Select Committee reexamined this evidence in 1977 and reconfirmed that the bullet and fragments had come from that Mannlicher-Carcano rifle.

The House panel employed a new form of neutron activation analysis (NAA) to match the recovered bullet and fragments to the ammunition used in the Mannlicher-Carcano. In this technique, traces from the ballistic evidence were bombarded by neutrons in a nuclear reactor so that the precise composition of elements—antimony, silver, and copper—could be measured by their emissions on a gamma-ray spectrometer to an accuracy of one-billionth of a gram. The composition of traces from the bullet and fragments were thus compared to that of the unfired bullet found in the chamber of the Mannlicher-Carcano and found to match exactly.

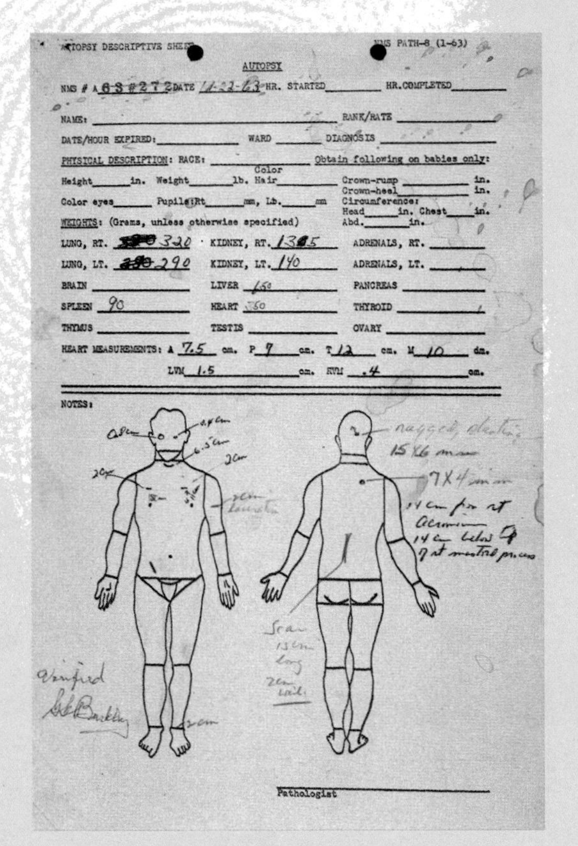

AUTOPSY DESCRIPTIVE SHEET   NMS PATH-8 (1-63)

AUTOPSY

NMS # A 63 #272 DATE 11-22-63 HR. STARTED ______ HR. COMPLETED ______

NAME: ______ RANK/RATE ______

DATE/HOUR EXPIRED: ______ WARD ______ DIAGNOSIS ______

PHYSICAL DESCRIPTION: RACE: ______ Obtain following on babies only:

Height ______ in. Weight ______ lb. Color Hair ______ Crown-rump ______ in. Crown-heel ______ in.

Color eyes ______ Pupils: Rt ______ mm, Lt. ______ mm Circumference: Head ______ in. Chest ______ in. Abd. ______ in.

WEIGHTS: (Grams, unless otherwise specified)

LUNG, RT. 320 · KIDNEY, RT. 13_5 ADRENALS, RT. ______

LUNG, LT. 290 KIDNEY, LT. 140 ADRENALS, LT. ______

BRAIN ______ LIVER 650 PANCREAS ______

SPLEEN 90 HEART 350 THYROID ______

THYMUS ______ TESTIS ______ OVARY ______

HEART MEASUREMENTS: A 7.5 cm. P 7 cm. T 12 cm. M 10 dm.

LVM 1.5 cm. RVM .4 cm.

NOTES:

15 X 6 mm

7 X 4 mm

Verified

Pathologist

A page from the autopsy report: "There is a large irregular defect of the scalp and skull . . . an actual absence of scalp and bone producing a defect which measures approximately 13 cm. in greatest diameter."

This analysis showed that all the ballistic material that was recovered, and could be tested, came from two bullets, and both bullets identically matched in their composition the ammunition for the Mannlicher-Carcano rifle.

Regarding Oswald's alleged murder of Officer Tippit, the FBI found that cartridge cases found at the murder scene matched the firing pin of the revolver taken out of Oswald's hand when he was arrested. The FBI determined no other weapon could have ejected these cartridges—and the House Committee's firearms panel agreed.

In addition, five eyewitnesses identified Oswald from the police lineup as either the person who shot Tippit or the person who fled from the scene, pistol in hand.

QUESTIONED DOCUMENTS

Other investigations by the Warren Commission and the House Committee also linked the JFK murder weapon to Oswald. They indicated that the person who ordered the rifle under the name "A. Hidell" from a mail-order house in Chicago in March 1963 and rented the post office box in Dallas to which it was shipped was Oswald. Numerous government handwriting experts determined in 1964 that Lee Harvey Oswald had signed the name "A. Hidell" on both the purchase order for the rifle and the post box application. A half dozen other documents found in Oswald's possession also indicated he had used the alias "Hidell." The House Select Committee panel of questioned document experts, after re-examining the signatures, unequivocally agreed that Oswald had ordered the murder weapon through his post office box.

Oswald was also photographed holding the rifle on March 31, 1963, and five witnesses including his wife testified that he had such a rifle in his possession. Although some conspiracy theorists have argued that the photo was doctored, most experts (and Marina Oswald, who said she took the picture) have defended its authenticity. Mrs. Oswald further stated in her affidavit: "Two weeks ago I was in the garage and saw the same blanket the police got. I opened the blanket and and saw a rifle in it. This blanket is the same one I saw today in the same place."

FINDINGS OF THE HOUSE SELECT COMMITTEE ON ASSASSINATIONS IN THE ASSASSINATION OF PRESIDENT JOHN F. KENNEDY (1978)

A. Lee Harvey Oswald fired three shots at President John F. Kennedy. The second and third shots he fired struck the President. The third shot he fired killed the President.

1. President Kennedy was struck by two rifle shots fired from behind him.

2. The shots that struck President Kennedy from behind him were fired from the sixth-floor window of the southeast corner of the Texas School Book Depository building.

3. Lee Harvey Oswald owned the rifle that was used to fire the shots from the sixth-floor window of the southeast corner of the Texas School Book Depository building.

4. Lee Harvey Oswald, shortly before the assassination, had access to and was present on the sixth floor of the Texas School Book Depository building.

5. Lee Harvey Oswald's other actions tend to support the conclusion that he assassinated President Kennedy.

B. Scientific acoustical evidence establishes a high probability that two gunmen fired at President John F. Kennedy. Other scientific evidence does not preclude the possibility of two gunmen firing at the President. Scientific evidence negates some specific conspiracy allegations.

C. The committee believes, on the basis of the evidence available to it, that President John F. Kennedy was probably assassinated as a result of a conspiracy. The committee is unable to identify the other gunman or the extent of the conspiracy.

1. The committee believes, on the basis of the evidence available to it, that the Soviet Government was not involved in the assassination of President Kennedy.

2. The committee believes, on the basis of the evidence available to it, that the Cuban Government was not involved in the assassination of President Kennedy.

3. The committee believes, on the basis of the evidence available to it, that anti-Castro

THE SCIENTISTS

Michael M. Baden, M.D.—one of the world's leading forensic pathologists, former longtime Chief Medical Examiner of New York City, co-director of the New York State Police Forensic Services Unit; best-selling author and seen in popular television programs such as HBO's "Autopsy" series. "Dr. Death" has worked on hundreds of high-profile cases including O.J. Simpson, the JFK and Martin Luther King Jr. assassinations, the Attica prison riot, and many more.

Cuban groups, as groups, were not involved in the assassination of President Kennedy, but that the available evidence does not preclude the possibility that individual members may have been involved. The committee believes, on the basis of the evidence available to it, that the national syndicate of organized crime, as a group, was not involved in the assassination of President Kennedy, but that the available evidence does not preclude the possibility that individual members may have been involved.

4. The Secret Service, Federal Bureau of Investigation and Central Intelligence Agency were not involved in the assassination of President Kennedy.

D. Agencies and departments of the U.S. Government performed with varying degrees of competency in the fulfillment of their duties. President Kennedy did not receive adequate protection. A thorough and reliable investigation into the responsibility of Lee Harvey Oswald for the assassination of President Kennedy was conducted. The investigation into the possibility of conspiracy in the assassination was inadequate. The conclusions of the investigations were arrived at in good faith, but presented in a fashion that was too definitive.

Polaroid taken by Mary Moorman as Kennedy is shot in the head. Her snap captures the grassy knoll. Is there a hatted man behind the picket fence? Mary Moorman was not asked to testify before the Warren Commission.

1. The Secret Service was deficient in the performance of its duties.

(a) The Secret Service possessed information that was not properly analyzed, investigated or used by the Secret Service in connection with the President's trip to Dallas; in addition, Secret Service agents in the motorcade were inadequately prepared to protect the President from a sniper.

(b) The responsibility of the Secret Service to investigate the assassination was terminated when the Federal Bureau of Investigation assumed primary investigative responsibility.

2. The Department of Justice failed to exercise initiative in supervising and directing the investigation by the Federal Bureau of Investigation of the assassination.

3. The Federal Bureau of Investigation performed with varying degrees of competency in the fulfillment of its duties.

(a) The Federal Bureau of Investigation adequately investigated Lee Harvey Oswald prior to the assassination and properly evaluated the evidence it possessed to assess his potential to endanger the public safety in a national emergency.

(b) The Federal Bureau of Investigation conducted a thorough and professional investigation into the responsibility of Lee Harvey Oswald for the assassination.

(c) The Federal Bureau of Investigation failed to investigate adequately the possibility of a conspiracy to assassinate the President.

(d) The Federal Bureau of Investigation was deficient in its sharing of information with other agencies and departments.

4. The Central Intelligence Agency was deficient in its collection and sharing of information both prior to and subsequent to the assassination.

5. The Warren Commission performed with varying degrees of competency in the fulfillment of its duties.

(a) The Warren Commission conducted a thorough and professional investigation into the responsibility of Lee Harvey Oswald for the assassination.

(b) The Warren Commission failed to investigate adequately the possibility of a conspiracy to assassinate the President. This deficiency was attributable in part to the failure of the Commission to receive all the relevant information that was in the possession of other agencies and departments of the Government.

(c) The Warren Commission arrived at its conclusions, based on the evidence available to it, in good faith.

(d) The Warren Commission presented the conclusions in its report in a fashion that was too definitive.

Ballistic Fingerprinting

A system to link firearms evidence such as bullets and cartridge casings found at a crime scene to a specific weapon using an automated search of a large database (ballistic fingerprinting) is technically achievable but politically controversial. Calls for the establishment of this crime-fighting tool were made in the wake of the case of the D.C. Snipers who terrorized the Washington-Maryland area in late 2002. In that case, police rapidly matched bullet fragments from each victim to prove that the same gun was used in all of the shootings, but they were unable to pinpoint who had bought that weapon because there is no nationwide database of ballistic fingerprints for every gun sold. Fierce opposition from the gun lobby and the National Rifle Association (NRA) has blocked efforts to institute the practice.

Whenever a gun is fired, identifying marks are made on the bullets and cartridge casings. These marks, called ballistic fingerprints, are as unique as human fingerprints.

Federal law enforcement agents have been using ballistic fingerprinting systems to match bullets to crime guns for more than a decade. In 1999, the FBI and the Bureau of Alcohol, Tobacco and Firearms (BATF) combined their ballistic fingerprinting efforts into a coordinated law enforcement system known as NIBIN (National Integrated Ballistics Information Network) that is now deployed in more than 160 different local law enforcement sites around the country. Law enforcement officials have been entering ballistic fingerprinting images from crime guns and bullet fragments into the databases, but gun manufacturers have refused to cooperate.

Some state legislatures have adopted the so-called "melting-point" laws designed to remove Saturday Night Specials from the market. These laws prohibit the sale of handguns with a barrel, slide, frame or receiver that will melt under a certain temperature. One of the justifications is that such guns do not retain ballistic "fingerprints" because irregularities are "burned off." But cheaper guns actually have *more* irregularities and are not fired often enough to lose them.

THE SCIENTISTS

Cyril H. Wecht, M.D., J.D.—forensic pathologist, former chief pathologist and coroner of Allegheny County, Pennsylvania, Dr. Wecht holds numerous academic and professional appointments in forensic science. He has served as an expert witness in many high-profile cases including Mary Jo Kopechne, Sunny von Bulow, Jean Harris, Dr. Jeffrey McDonald, the Waco Branch Davidian fire, and Vincent Foster.

Ballistics, Handwritting Analysis, Profiling: Son of Sam, 1976–77

A psychopath terrorizes New York with a series of random attacks.

From 1976 to 1977, a crazed killer dubbed "Son of Sam" murdered six young people and wounded several others in seemingly random attacks in New York City. What inspired the most terror were his bizarre, scrawled letters to victims, the police, and a popular newspaper columnist, in which he bared his tortured soul. Before he was caught, some observers came up with a new term to describe him and others like him. They called him a "serial killer."

It started at one o'clock in the morning on July 29, 1976, as two young women, Jody Valenti and Donna Lauria, were talking in a car outside their Bronx apartment when a strange man suddenly stuck his head in the window and unleashed five deafening blasts from a handgun. Valenti was hit in the thigh, Lauria gushed blood from a fatal wound to her neck.

Three months later, a young man was shot in the neck as he sat in his car; a month after that, two girls were wounded in their parked car. All three survived, but one victim was paralyzed for life. Two months later, Christine Freund was murdered as she sat in her car with her boyfriend near the Forest Hills train station. It wasn't until this fatal attack that police were able to recover bullets from the crime scene that suggested a thrill killer, using a .44-caliber pistol, was at work.

Then, in March 1977, the mystery assailant shot and killed Virginia Voskerichian, prompting the New York Police Department to announce it was establishing a special unit with 100 detectives, Operation Omega, headed by Captain Joseph Borelli, to catch the killer. But that didn't deter the gunman. Alexander Esau and Valentina Suriani were the next to lose their lives.

A rambling hand-written letter, signed "Son of Sam," was later found at the crime scene. Full of misspellings, it was addressed to Captain Borelli, the commander who had vowed to track him down. The killer wrote:

> I am deeply hurt by your calling me a weman-hater. I am not. But I am a monster. I am the "son of Sam." I am a little brat. When father Sam gets drunk he gets mean. He beats our family. Sometimes he ties me up to the back of the house. Other times he locks me in the garage. Sam loves to drink blood. "Go out and kills" commands father Sam. Behind our house some rest. Mostly young—raped and slaughtered—their blood drained—just bones now. Pap Sam keeps me locked in the attic too. I can't get out but I look out the attick window and watch the world go by. I feel like an outsider. I am on a different wavelength then everybody else—programmed to kill. However, to stop me you must kill me. Attention all police: shoot me first—shoot to kill or else keep out of my way or you will die. Papa Sam is old now. He needs some blood to preserve his youth. He has too many heart attacks. "Ugh, me hoot, it hurts, sonny boy." I miss my pretty princess most of

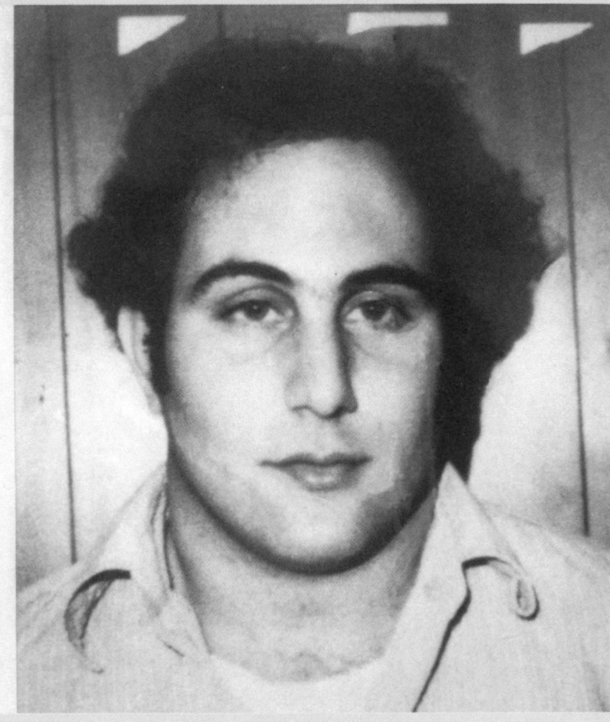

Police sketch and the official police photograph of David Berkowitz. Son of Sam got religion in prison: “Because of Jesus Christ and my faith in Him I am trying to make amends to society in any way that I can.” (March 25, 2002, just before a parole hearing.)

> all. She’s resting in our ladies house. But i’ll see her soon. I am the “monster”—“Beelzebub”—the chubby behemouth. I love to hunt. Prowlling the streets looking for fair game—tasty meat. The wemon of Queens are Z prettyist of all. I must be the water they drink. I live for the hunt—my life. Blood for papa. Mr. Borelli, sir, I don't want to kill any more. No sur, no more but I must, “honour thy father.” I want to make love to the world. I love people. I don't belong on earth. Return me to yahoos. To the people of Queens, I love you. And i want to wish all of you a happy Easter. May God bless you in this life and in the next. And for now I say goodbye and goodnight. Police: Let me haunt you with these words: I'll be back. I'll be back. To be interrpreted as—bang, bang, bang, bang—ugh. Yours in murder, Mr. Monster.

The letter contained many clues and insights about the killer, but by the time it reached Borelli it had been handled by so many people that his forensic specialists couldn’t find any telltale fingerprints or other evidence to enable the police to find their suspect quickly. Handwriting specialists and criminal profilers analyzed it as best they could. When leaked to the news media, the letter created a wave of terror.

At the end of April, a similar letter was sent to Jimmy Breslin, a workingman's columnist at the *New York Daily News* who had been closely covering the story. When Breslin wrote about his correspondence, the paper's circulation hit a sales record. Everybody was afraid of Son of Sam.

By now police had identified the weapon involved as a .44-caliber Charter Arms Bulldog revolver and detectives tried to trace every one of the 28,000 guns of that type that had ever been manufactured, including as many as they could of the 600 guns reported stolen.

A hundred more detectives were assigned to Operation Omega. Forensic psychiatrists analyzed the killer's letters and diagnosed him as a paranoid schizophrenic. Others pointed out that the attacks seem to have occurred just before or after a full moon. Desperate for any lead they could find, the detectives listened patiently as astrologers, seers, numerologists, and shrinks came forward to pitch their theories about Son of Sam.

But the attacks continued. On June 25,1977, another couple were gravely wounded but survived. Then on July 30, 1977, in Brooklyn, Robert Violante was permanently blinded and his companion Stacy Moskowitz was murdered. This time a witness claimed to have glimpsed the shooter well enough to help police create a composite sketch. They went on the lookout for a white male between 20 and 35 years old, 5 feet 7 inches to 6 feet 2 inches tall, 150 to 220 pounds. Somebody said he had escaped in a yellow Volkswagen. All of his victims had been shot at point-blank range. The fact that he fired his weapon from a combat stance led investigators to believe he had received military firearms training. Detectives scrambled to check the backgrounds of anyone who appeared to fit the bill. At one point they had as many as 3,000 suspects.

At the last crime scene, a witness claimed to have seen a man removing a parking ticket from his windshield. Police ran down a car that had received a ticket for being parked too close to a fire hydrant near the site of the last attack.

The suspect was David Berkowitz, 23, an Army veteran and sometime security guard.

On August 10, 1977, they found the cream-colored Ford Galaxy parked near the owner's home at 35 Pine Street in Yonkers and kept it under surveillance. As a stocky young man got into the car, a squad from Operation Omega swooped down and confronted him. Berkowitz responded by saying, "Well, you got me," and when asked who he was, he said, "Sam."

Cops found a machine gun protruding from a sack in his car. Inside his apartment they found the walls covered with eerie slogans such as "KILL FOR MY MASTER," and a diary that documented each of the 1,488 fires he had set throughout the city. Berkowitz quickly confessed, saying he received telepathic commands to kill from a 3,000-year old demon who lived inside a dog named Harvey, which belonged to his neighbor, Sam Carr. He was found competent to stand trial and was ultimately convicted and sentenced to 365 years in prison.

In an era when computer and laboratory capabilities were extremely limited, the NYPD utilized many methods to try to catch Son of Sam. But in the end it was an observant eyewitness account from a crime scene, combined with thorough, old-fashioned "shoe-leather" detective work and an alert and informed public, which nabbed the serial killer.

Blood Patterns

Efforts have continued for decades to identify the source of blood found at a crime scene. In 1939, a German scientist, Victor Balthazard, suggested that police could use bloodstain patterns to help them determine the dynamics of a crime. By the early 1950s, a science of blood-spatter pattern analysis began to arise in the U.S. One of its leading pioneers, Dr. Paul Kirk of the University of California at Berkeley, urged closer police attention to the shape and location of bloodstain patterns, after he pointed out that they indicated the volume of blood, the speed at which the drops had fallen, the direction or angle of motion, angle of impact, and other facts that could help reconstruct a crime scene.

Blood Transfer

Excerpt of trial testimony of criminalist Herbert Leon MacDonnell, questioned by the defense about blood patterns in the O.J. Simpson murder case:

> **Q.** How did you make a determination as to how that blood was applied to that sock, from your analysis, looking at it through a microscope?
> **A.** Well, more or less, by elimination, I determined how it didn't get there . . . We can eliminate those kinds of stains that would produce other results. For example, blood did not drip onto this area; it did not splash or spatter onto it. It was transferred by one of two mechanisms which are very closely associated: one would be simply touching or compressing it; and another would be a lateral motion at the same time, which is called a swiping action, as differentiated from wiping, where you wipe something up and the stain is already there, like on a countertop, if you wipe it up. But if you have blood, for example, on your finger, and you touch something with or without a lateral motion, it is called transfer. If there is a lateral motion, you may see some feathering out as it moves along and leaves the surface. And these edges were quite crisp. And while it could have been a swiping-type action, it is also consistent with a—just straight compression. And that could have resulted by either coming in contact with something that had blood on it or blood simply being added to the surface with something like a pipette or medicine dropper, or just gently putting it on so it didn't drop any distance, or it would have caused satellite spatters, I've seen other spots around it. So this is just a transfer pattern, either by something like a finger that's very, very bloody, touching in a perfect oval, which is not logical but possible, or a drop of blood, a single drop of blood that is added and "teased around," more or less moved, to create a stain to soak into the fabric.

Blood Spatter: Sheppard

A brilliant criminalist devises new techniques to follow the killer's bloody trail—and helps to free Sam Sheppard.

The blood was everywhere. Police arrived at the physician's Bay Village, Ohio, home early on July 4, 1954, to find Mrs. Marilyn Sheppard in the upstairs master bedroom. Her body was lying in her bed in a pool of blood, her head battered by more than fifteen blows from an unidentified heavy object. A pillow on the bed showed a three-inch-long bloodstain. The walls of the master bedroom were spattered with blood. A trail of blood led down the stairs and onto the terrace.

Dr. Sam Sheppard, husband of the deceased, his face swollen and one of his vertebrae fractured, seemed disoriented from his ordeal. But there was no cut or blood spots on him.

He told the cops a harrowing story about how the previous night his wife's screams had wakened him from his sleep on the downstairs daybed. He had rushed upstairs and entered the bedroom where he was struck on the head from behind. When he came to, he found his wife, dead, and panicked about their son, Sam, aged seven. Then he realized the intruder was still in the house, so he gave chase onto the beach along Lake Erie. He caught a man and fought him, but the man knocked him out again. He described the stranger as tall with a large head and bushy hair.

The detectives didn't believe Sheppard's story, especially when they caught him in a lie. Upon learning that he had tried to conceal from them his adulterous affair with a medical technician, they considered him the prime suspect. Word went out that Sheppard had cheated on his pregnant wife and then killed her during a fight over his infidelity.

The *Cleveland Press* covered the story with a fury, openly accusing Sheppard of murder in a barrage of editorials. Three weeks after the murder, Sheppard was arrested and charged with the crime. His trial turned out to be the worst media circus since the Lindbergh kidnapping. Everybody was posturing for the press. Somebody said the prosecution had charged him with murder, but proved only adultery, yet he was convicted. As the judge sentenced Sheppard to life imprisonment, he kept insisting he was innocent.

A dramatic moment in Cuyahoga County Court, at Sam Sheppard's preliminary hearing. His brother Dr. Richard whispers to him—what?

Shortly after they hauled him off to prison, his mother killed herself and his father died. He considered suicide, but decided he had to prove to his son, their son, Sam, that he was innocent. Sheppard vowed to find the real murderer.

DR. KIRK

As Sheppard's case went up for an appeal, his lawyer hired one of the nation's leading criminalists, Dr. Paul Leland Kirk of the University of California at Berkeley, to help them assess the evidence in order to reconstruct the murder. Kirk, an accomplished microchemist who had worked on the Manhattan Project that built the first atomic bomb, had later switched to criminalistics and recently published the leading textbook in that field. Kirk warned that he would conduct an independent and objective study. Then he started his analysis.

He was immediately shocked to find how badly the police had bungled much of the blood evidence from the crime scene. The authorities had minutely recorded hundreds of bloodstain patterns on the bedroom wall, but not attempted to analyze them. Neither had they tested the bloody trail leading from the bedroom through the living room and out on the terrace, nor had they grouped any of that blood. Kirk scoured the Sheppard home. He took blood samples from the bedroom walls. He vacuumed the carpets with a special sweeper equipped with a custom-made filter to trap minute particles. He removed various other samples from the house and grounds.

Rummaging around in the Sheppard house crime scene, Dr. Samuel R. Gerber had been elected coroner of Cuyahoga County in 1936 and knew Dr. Sam Sheppard from local medical circles. He later convened an inquest on network TV.

Having found no blood traces on the bedroom ceiling, Kirk reasoned that the murder weapon had been wielded in a more or less horizontal fashion that had kept the blood from flying up. This was evident from the blood splashes on the walls, some of which must have been flung from the murder weapon as it was swung backward and forward to make contact with the victim's head. Other spatters had come directly from the woman's head as it was being battered.

Based on the blood impression left on the pillow and the nature of the wounds to the victim's head and other factors, Kirk's experiments led him to conclude that the most likely murder weapon had been a blunt object, not a surgical instrument as the police claimed. Noting the blood drops that had been smeared into streaks on the right side of the victim's bed, Kirk deduced that the murderer had landed his blows while standing between the twin beds. The blood-free areas on two of the walls behind led suggested the murderer's own body had blocked blood from splattering there.

Some of this information worked to Dr. Sheppard's advantage. A killer standing in that position must have swung the murder weapon with his left hand, but Dr. Sheppard was neither left-handed nor ambidextrous. Furthermore, the crime scene indicated that the murderer must have been thoroughly spattered with blood, yet Sheppard had been found to have no blood on his clothes, apart from a bloodstain on his trouser knee, which he said had occurred when he had knelt by the bed to take his wife's pulse. After four months of evidence gathering and analysis, Kirk completed a 10,000-word affidavit. Some of his amazing findings included the following:

- The absence of blood on the ceiling indicated that the blows were made from horizontal, not upward, swings.
- The murder weapon was likely to have been less than a foot long, cylindrical in shape and relatively lightweight, like a metal flashlight. He noted that red lacquer paint that was found at the scene could have come from the murder weapon rather than from nail polish, since Marilyn was not wearing any fingernail or toenail polish at the time of her death. Kirk proved that it was in fact commercial lacquer used to paint hardware and metal.
- Two fragments of Marilyn's upper front teeth had been found on the bed, but the autopsy did not detect any injuries to her face, indicating that she likely broke them when she bit her attacker. Dr. Sheppard did not have any such wounds.
- A large spot of blood found in the murder bedroom matched neither Dr. Sam's nor Marilyn's type. Kirk concluded it belonged to the murderer.
- Marilyn's pajama pants had blood accumulated at the bottom, showing that they had been pulled down before the murder and, therefore, Kirk theorized that the crime had started as a sex attack.
- The three-and-one-half-inch tear on the right pocket of Dr. Sheppard's pants extended directly downward from the pocket, indicating that somebody else had ripped the clothing, which was consistent with Sheppard's account.
- Sand found in Sheppard's pants pocket could only have gotten there by his lying in the water for at least an hour.

After the state courts rejected the appeal without even considering Kirk's affidavit, Sheppard's lawyer appealed the case to the U.S. Supreme Court, arguing in his petition for a writ of certiorari that Sheppard had been denied his right to a fair trial. (A writ of certiorari—Latin for "to be ascertained"—when granted means that a higher court will examine the decision of a lower court for legal error.) Although the Court declined to review the case, Judge Felix Frankfurter's memorandum indicated that at least one justice thought that Sheppard's trial had been conducted like "a Roman holiday." When the Supreme Court denies such petitions, this explicitly does not mean that it has approved the lower court's decision.

Everything changed when Sheppard's family hired F. Lee Bailey, a brash young Boston lawyer. Bailey filed a habeas corpus petition in federal court. District Judge Carl A. Weinman of the United States District Court for the Southern District of Ohio considered Bailey's arguments and ruled that Sheppard's 1954 conviction was a "mockery of justice" that violated his constitutional rights in five ways. Weinman called the Sheppard case a classic example of trial by newspaper. He ordered Sheppard released immediately on a $10,000 bond, pending trial.

Kirk testified effectively at Sheppard's second trial in 1966. The three-week proceeding ended with Sheppard's acquittal. But out of court, county officials continued to claim he was guilty. After the verdict, Sheppard sank into alcohol and drug abuse, ultimately becoming a professional wrestler. He died, a broken man, of liver disease at age 46. But even then, the case didn't end.

SAM REESE SHEPPARD CONTINUES THE FIGHT

In 1998 Sheppard's son, Sam Reese Sheppard, sued the state of Ohio for $2 million for his father's wrongful imprisonment. He hired some top lawyers and forensics experts to try to clear his father's name and build a case against the man he considered his mother's real killer. His reexamination featured new scientific techniques including DNA that hadn't been available during the first two trials. But the state of Ohio waged a fierce battle in the same county where the first trials had occurred, determined to uphold its "good name."

Based on new evidence, Sheppard's attorney, Terry Gilbert, identified the real killer as Richard Eberling, a known criminal who had worked as a window washer at the Sheppard's house before the murder. Gilbert claimed Eberling had been arrested in 1959 with a ring of Marilyn Sheppard's in his possession; Eberling had oddly volunteered to police that he had once cut himself and bled at the Sheppard home some time before the murder. Sam Reese Sheppard said he had interviewed Eberling in prison and come away convinced that he was involved in the murder. Eberling reportedly had confessed to another inmate that he had committed the Sheppard murder. But Eberling, a convicted murderer, had died in prison in 1998.

Sheppard's team reconstructed a full-scale version of the family's former home and furnished it down to the minutest detail, exactly as it had appeared in 1954. Their blood expert, Bart Epstein of Minnesota, who had studied under Paul Kirk and gained access to his original notes, set out to recreate and restudy Kirk's landmark analysis. To conduct his blood spatter experiments, Epstein drew some of his own blood and used a specially constructed device to replicate the original blood patterns. His analysis upheld many of Kirk's original findings.

Dr. Cyril Wecht, the famed former Pittsburgh medical examiner, testified that the victim's wounds were consistent with "blunt-force trauma" marks that would have been left by a 1950s-era metal flashlight—not from a sharp surgical instrument as the prosecution had claimed. Wecht also agreed with Kirk's findings about the sexual assault. The legal team used Dr. Mohammad Ashraf Tahir, a forensic DNA specialist in Indianapolis, to attempt to test some of the surviving crime scene evidence. Tahir received a bloodstained wood chip that had been retrieved from the stairs of the Sheppard house, a section of blood-splattered flooring from the porch and a scraping from the closet door blood stain. He obtained Marilyn's DNA from a few strands of her hair collected after the murder. To get Dr. Sheppard's DNA, Sam Reese Sheppard granted permission to have his father's body exhumed for tissue sampling. Tahir also had a small sample of Eberling's blood available for DNA testing.

Tahir used a DQA1 test to assess the samples. He poured the DNA from each of the samples over a sensitized paper test strip that was printed with a set of numbers. Each number referred to the eight specific alleles the test can detect. Indicators turn blue when the test has detected a specific allele. A test of the wood chip found on the stairs, and the blood stain lifted from the porch, which the prosecution had contended had contained the victim's blood, did not show Marilyn or Samuel Sheppard's alleles. However, Richard's Eberling's allele matched one of the stains on the blood trail, and therefore, Eberling could not be excluded. As Tahir predicted, the sample from the largest bloodstain, the spot on the closet door, appeared to be a perfect mixture of Eberling's blood and Mrs. Sheppard's.

But the DNA evidence against Eberling would not stand up to today's DNA evidence requirements. Prosecutors downplayed the validity of the DNA tests as "mumbo jumbo," saying the entire crime scene had been contaminated, as in the O.J. Simpson case. They insisted that Marilyn Sheppard's damaged lower teeth proved she had been struck a hard blow to her mouth, not that she had broken a tooth from biting her assailant, pointing out that Sam Sheppard was a boxer capable of landing such a blow. In 2000 the civil court jury finally reached its verdict. It decided that Sheppard's son had failed to meet the statute's burden of proof in his wrongful imprisonment lawsuit. Once again, outstanding forensic evidence had failed to "clear" Dr. Sam Sheppard in the local community's eyes.

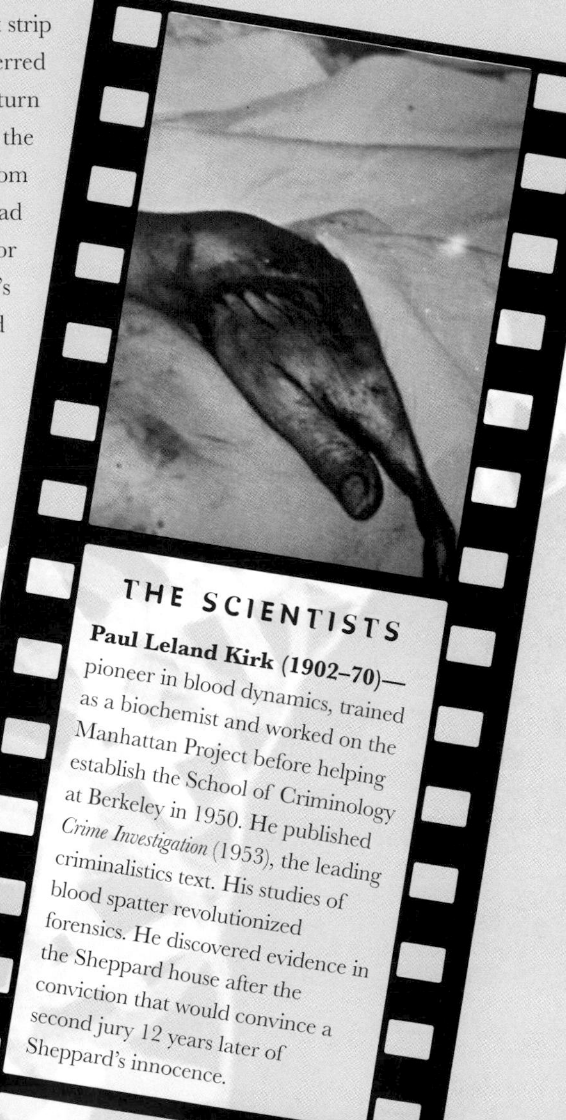

THE SCIENTISTS

Paul Leland Kirk (1902–70)—pioneer in blood dynamics, trained as a biochemist and worked on the Manhattan Project before helping establish the School of Criminology at Berkeley in 1950. He published *Crime Investigation* (1953), the leading criminalistics text. His studies of blood spatter revolutionized forensics. He discovered evidence in the Sheppard house after the conviction that would convince a second jury 12 years later of Sheppard's innocence.

Head Trips and Reading the Signs

Some companies use psychological profiling to tailor their products and workforce. They mine any available data, from public records to blogs. The criminalist may use similar methods in an effort to identify an unknown person. There have been some remarkably accurate criminal profiles, such as that of Russian serial killer Andrei Chikatilo, right. (Chikatilo was so impressed by psychiatrist Dr. Alexander Bukhanovsky's profile that he asked the doctor to attend his execution.) Others have been positively misleading. Similarly, forensic investigation of documents has had some successes and some highly publicized failures.

Dr. Bukhanovsky accurately predicted that Chikatilo would be an impotent heterosexual with an exceptional memory.

The unknown "Zodiac" killer sent a code to San Francisco newspapers in 1969. Despite all his written communications, he was never caught. He killed at least five, though his letters suggested 37. He threatened to go on a "kill rampage" if the papers did not publish the messages. His random and motiveless killings made it impossible to construct patterns that would help lead to him.

Psychological Profiling: George Metesky

Before the Unabomber there was the Mad Bomber. He wasn't mad enough to post letters from his hometown, but the profiler assumed that.

Starting on November 16, 1940, and lasting for sixteen years, New York City remained in the grip of a mysterious terrorist whom the media dubbed the "Mad Bomber." Finally, it was the police who sought psychiatric help. The break came when the psychiatrist gave the cops a detailed criminal profile that enabled them to get their man.

At first, police scarcely noticed the crudely built pipe bomb and threatening note that had been discovered outside a Consolidated Edison office in Manhattan. It wasn't considered a big deal. A second, similar device turned up a year later, but it too was dismantled before it went off. Still not such a big deal in New York. In December 1941, however, shortly after the Pearl Harbor attack drew the United States into World War II, the NYPD received a bizarre note that had been written in crude block letters and mailed anonymously from Westchester County.

> "I WILL MAKE NO MORE BOMB UNITS FOR THE DURATION OF THE WAR—MY PATRIOTIC FEELINGS HAVE MADE ME DECIDE THIS—LATER I WILL BRING THE CON EDISON TO JUSTICE—THEY WILL PAY FOR THEIR DASTARDLY DEEDS . . . F.P."

Now the police were taking closer note. Based on the latest clue, they called their prey "F.P." After the war, the Mad Bomber struck again. An undetonated explosive was uncovered on March 29, 1950, at Grand Central Station. Another blew up in a public phone booth outside the New York Public Library. Still more bombs showed up, strapped to movie theater seats and other horribly vulnerable places—often frighteningly close to going off. Eventually, in 1953, some of the bombs became more sophisticated and they began to cause some serious injuries.

As the Mad Bomber taunted his pursuers with letters and phone calls, the newspapers demanded action. Then, after a Christmas-season 1956 blast injured six people in a Brooklyn theater, the police became desperate.

Psychiatric profiling had been employed for several years for military intelligence purposes, but nobody had yet developed criminal profiling. However, NYPD Inspector Howard E. Finney heard about a psychiatrist with a reputation for being unusually insightful, so Finney approached Dr. James A. Brussel with clues about the bomber and the doctor began his analysis.

Brussel carefully studied the letters and other evidence before formulating his conclusions. Using his powers of induction and deduction, he said the neat printing suggested a meticulous planner who took pride in his work. Such qualities meant the bomber had likely been a model employee. His messages indicated he probably had received a higher-than-average education, but lacked common American slang and idioms. He appeared foreign-born, and based on his language structure was probably a man of Slavic descent and most likely Roman Catholic.

Noting that the envelopes of all letters were postmarked in Westchester County, Brussel said F.P. was too clever to mail them from his hometown. He determined that the largest congregation of Slavs in the Tri-State area was in Connecticut. Bridgeport had a very large Polish community, and the route to Manhattan leads through Westchester County. Therefore Brussel deduced that F.P. possibly lived in southern Connecticut.

Based on the handwriting, especially the curved W's shaped like breasts, and other clues, the doctor assumed an Oedipal complex. From his psychiatric experience he suspected the bomber's mother was probably deceased and that he lived with an older female relative, probably a sister. He thought the bomber was a social loner who craved attention.

Brussel urged the police to publicize this profile, predicting it would prompt the bomber to respond in a way that would reveal more about himself. Brussel offered one more deduction. He said when the bomber was found, he would be wearing a double-breasted suit. Buttoned.

"Mad Bomber" George Metesky living up to his nickname in Waterbury Jail, Connecticut, January 21, 1957, shortly after his arrest. Dr. James Brussel's path-breaking psychological profile proved to be fantastically accurate.

Case Study: George Metesky

The police followed Brussel's recommendations. The public release of the profile triggered a deluge of false confessions and lousy leads. But then Brussel received an agitated phone call from the bomber. The caller's warning to "stay away" suggested to Brussel that his deductions about the bomber were relatively accurate.

Then another lead surfaced. The bomber had written to the *Journal American* noting the date he had suffered injuries that he blamed on "the Consolidated Edison." Based on the profile, Consolidated Edison had searched its records and found a former employee, George Metesky, who had worked for the company from 1929 to 1931, until being denied a disability pension. Some of his letters in the files contained similar phraseology to that used by F.P. and he appeared to match closely the profile developed by Brussel. Now investigators had a possible motive.

Finney and his detectives arrived at Metesky's Connecticut residence late at night. Metesky answered the door in his pajamas. In appearance and manner he matched the profile. To their surprise, he confessed on the spot and proudly showed off his bomb-making workshop.

The detectives asked him to get dressed for a trip to the police station. Metesky politely obliged. He came down wearing a pinstriped double-breasted suit. Buttoned.

After his arrest, Metesky was found insane and was committed to Matteawan State Hospital for the Criminally Insane, where he remained until his release in 1973. Brussel refused to accept credit for the arrest, calling his work a blend of "science, intuition, and hope." He also warned that a psychological profiler's development of an after-the-fact typology of offender characteristics should never take the place of dogged police work based on hard evidence. In the 1960s he also devised a profile for the Boston Strangler that police later said fit the man they arrested, Albert DeSalvo. (DeSalvo was ultimately convicted of other offenses and some historians continue to question his guilt in the murders, or some of the murders.)

Brussel's successful analysis eventually gave rise to profiling specialties within some police departments, most notably the FBI's Behavioral Science Unit in the National Center for the Analysis of Violent Crime, which was formed in the early 1970s. It became popularized through Thomas Harris's novel, *Silence of the Lambs*, filmed in 1991.

The FBI released its Unabomber psychological profile in 1993, predicting that the killer would be extremely bright, probably a white man, highly educated and no longer young, with a passion for exactitude and a macabre sense of humor. In some respects, the profile proved right. But it also fit a lot of subjects. The Unabomber case was eventually solved by Ted Kaczynski's concerned brother, David, who contacted law enforcement authorities about his suspicions in the case, based on his reading of some of the Unabomber's published philosophical writings.

Despite the public perception of its unerring, uncanny exactitude, profiling efforts have often proved to be of limited value and some profiles may have actually diverted police attention in the wrong direction. This apparently happened in the cases of the Atlanta Olympics Bomber and the D.C. Area Sniper. Yet empirical studies have underwritten the efficacy of the technique.

Questioned Documents: Mark Hoffman

A series of deadly bombings gets traced to explosive documents involving the Mormon Church. And the end result did not cast document experts in a good light. Forensic investigation of a series of deadly bombings in Utah ended up cracking the case to reveal one of the great forgers in American history while at the same time casting doubt on the proficiency of most "experts" who detect questioned documents.

In October 1985 three package pipe bombs exploded in Utah, killing two persons and injuring a third. Police lacked a threatening note or other available forensic evidence to break the case. But investigators soon began to suspect the last victim after his account about how a strange package had gone off in his car turned out to be a lie. Instead of what he had described, he appeared to have been kneeling in the car tinkering with the package when it blew up. Police suspected he might be the bomber. But to prove he had committed both murders, they would need to find his motive.

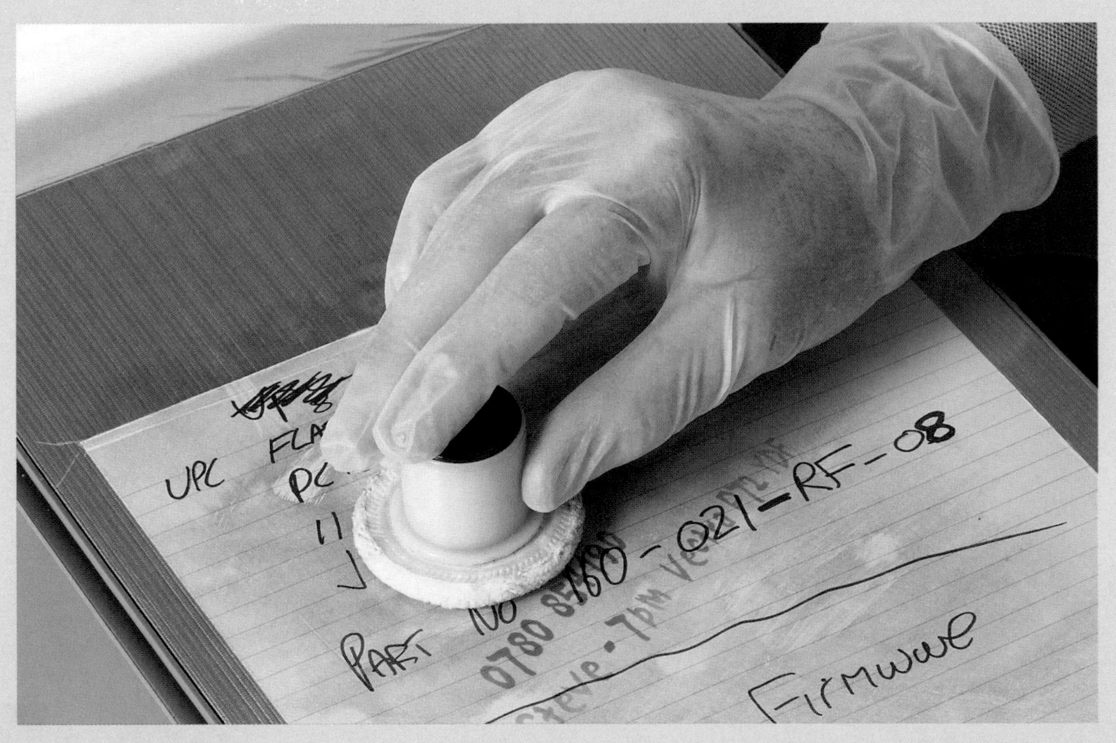

Electrostatic imaging system for detecting indented writing on documents; applying the toner pad. The soft brushed textile pad impregnated with toner is wiped across the imaging film. The toner is charge-sensitive and shows up the invisible electrostatic image.

Case Study: Mark Hoffman

The suspect, Mark Hoffman, was a dealer in antique documents, particularly Mormoniana. A sixth-generation Mormon himself, he appeared universally respected and upright, and dozens of top libraries and other collectors, including the American Antiquarian Society and the Library of Congress, all attested to the authenticity of his wares. Yet investigation revealed that both of the other victims had been linked to a complicated business scheme involving an unsecured $185,000 loan to Hoffman that had been made by elders of the Church of Jesus Christ of Latter-day Saints, the Mormons. The loan was in arrears.

At issue was a literary collection of antique letters that Hoffman had offered to sell to the church. The contents of the papers included something called the "white salamander letter," which appeared to refute some of the Mormons' basic teachings as laid down by the church's prophet and founder, Joseph Smith, in the Book of Mormon.

If it was authentic, the salamander letter threatened to refute the entire basis for Mormonism. Many believed that the Mormon Church hierarchy would not have wanted such secrets to leak out and some dealers speculated that Hoffman may have been targeted for attesting to its authenticity and agreeing to make it available to the Church. However, if the documents were forgeries, perhaps Hoffman had been trying to cover up his fraud.

During a search of Hoffman's van and premises, investigators uncovered some suspicious books about forgeries and bomb making. On the other hand, he passed a polygraph examination and FBI document experts continued to validate the authenticity of Hoffman's letters.

But Special Agent George J. Throckmorton, a devout Mormon, continued to doubt the authenticity of the salamander letter. Working with an experienced document examiner, William J. Flynn (who was not a Mormon), he began to study all of Hoffman's documents, not just the salamander letter. Using a book about forgeries they had obtained from Hoffman's home, Throckmorton and Flynn figured out the recipe and method Hofmann had used for making iron gall ink. Previous tests had not been able to determine the ink's age, but the pair were able to prove the documents were forgeries. One of the clues was the "alligator effect"—the microscopically detected cracking and breaking of the ink that had been caused by chemical oxidants that had been used to make the ink look older than it was.

Cutaway of a typical pipe bomb as constructed by Mark Hoffman. Though as such devices are homemade, they are necessarily unique, so any trace evidence recovered, such as wadding, nails, bolts. etc., can provide strong evidence.

Hoffman was bound over for trial on thirty felonies, including two capital murders. When confronted with the evidence against him, he pleaded guilty to reduced charges and was sentenced to one five-years-to-life term in prison. Hoffman ended up being hailed as one of the most brilliant forgers in history and the white salamander letter was shown to have been bogus, pleasing the Mormons. But the record of previous authentications by a long line of distinguished document examiners left nagging questions about the reliability of forgery detection.

Investigators assigned to bomb cases like to say that every bomb maker leaves his unique "signature" in the way he or she puts together the device. The Hoffman case involved questioned documents, within which forensics scientists attempt to find similarly unique characteristics.

Document examiners can face a variety of evidence that is produced by handwriting, copying machines, typewriters, cash register printers, facsimile machines or other sources. Careful study of the type of papers and inks used can try to pinpoint the source and potentially lead to identification of the purchaser, thereby exposing a fraud. Shortly after the Hoffman case, an article in the *University of Pennsylvania Law Review* criticized many of the techniques used in handwriting analysis as "junk science." By 1999, an influential report by the National Institute of Justice concluded: "Questioned document examination, which encompasses forgeries, tracings, and disguised handwritings, is currently in a state of upheaval." Under the standards the U.S. Supreme court laid down in *Daubert* (see page 18), one federal District Court decision in New Jersey in 2000 expressed concern about weaknesses in the reliability of handwriting analysis evidence, pointing out that there was: "No known error rate, no professional or academic degrees in the field, no meaningful peer review, and no agreement as to how many exemplars are required to establish the probability of authorship."

Another decision in Illinois expressed doubt about the ability of a document examiner to reach an accurate conclusion about a "match" between writings that a defendant is known to have authored and the writings in question, noting that therefore some judges would not allow the expert to express an opinion about authorship. Although science in the field may have made some advances, the legal bar has been raised.

GILBERT v. CALIFORNIA 388 U.S. 263 (1967)

A ruling in the above case means that a handwriting sample, in contrast to the content of what is written, like the voice or body itself, is an identifying physical characteristic outside the protection of the Fifth Amendment protection against self-incrimination. The defendant is not entitled to assistance of counsel during an exemplar session.

Questioned Documents/Profiling: Bioterrorist or Terrorists Unknown

The Anthrax Killer, who struck in the wake of the September 11 attack and terrorized Americans by sending deadly anthrax bacteria through the mails, has not been caught to date.

In the wake of the September 11, 2001, attack on the World Trade Center, crudely addressed envelopes began to arrive at major news outlets in New York and Boca Raton, Florida, and Washington, D.C., bearing a Trenton, New Jersey, postmark and containing a mysterious white powder. Some of the letters sat unopened in piles or mailbags, but others were opened. All of them were infectious.

On October 5, an editor at a Florida-based supermarket tabloid became the first known inhalation anthrax death in the U.S. since 1976 (or at least the first publicly attested). Anthrax was later found on his computer keyboard.

Attoney General John Ashcroft at Justice Department press conference, Washington, November 29, 2001. In the background is the wanted poster of fugitive Clayton Lee Wagner, self-styled anti-abortion warrior and one suspect (of many) in the anthrax bombings. "God called me to make war on his enemies and it does not matter to God or me if you're a nurse, receptionist, bookkeeper or janitor."

More potent than the first anthrax letters, the material found in letters sent to two United States Senators was a highly refined dry powder consisting of approximately one gram of nearly pure spores. Some reports described the material in the Senate letters as "weaponized" or "weapons grade" anthrax. This discovery caused the government to shut down its Capitol Hill mail service.

Exposure to anthrax killed five persons, seriously sickened 13, and caused hundreds more to undergo extensive medical treatment. But the greatest casualty was peace of mind for millions of Americans who feared the unknown, following an unprecedented germ warfare attack involving *Bacillus anthracis.* Suddenly ordinary American civilians were confronted with bioterrorism that could be delivered directly into their homes through the postal system.

In addition to killing one person at a targeted address in Florida, two other persons in New York and Connecticut died from unknown sources, possibly from cross-contamination of mail, and two other fatalities involved postal employees in the Brentwood facility in Washington. Investigators later found traces of anthrax in a street mailbox adjoining Princeton University, which caused them to suspect some of the infected letters had been mailed from that location. But they couldn't determine who had mailed the letters.

The anthrax attacks triggered a massive interagency investigation involving the FBI, the U.S. Army, the Environmental Protection Agency, the Centers for Disease Control, the U.S. Postal Service, and other law enforcement agencies. Laboratory testing of the materials used led investigators to discover that the letters contained at least two grades of anthrax material, but all of the anthrax was the same strain. This Ames strain had first been researched at the U.S. Army's Medical Research Institute of Infectious Diseases in Fort Detrick, Maryland, and later distributed to at least fifteen bioresearch labs within the

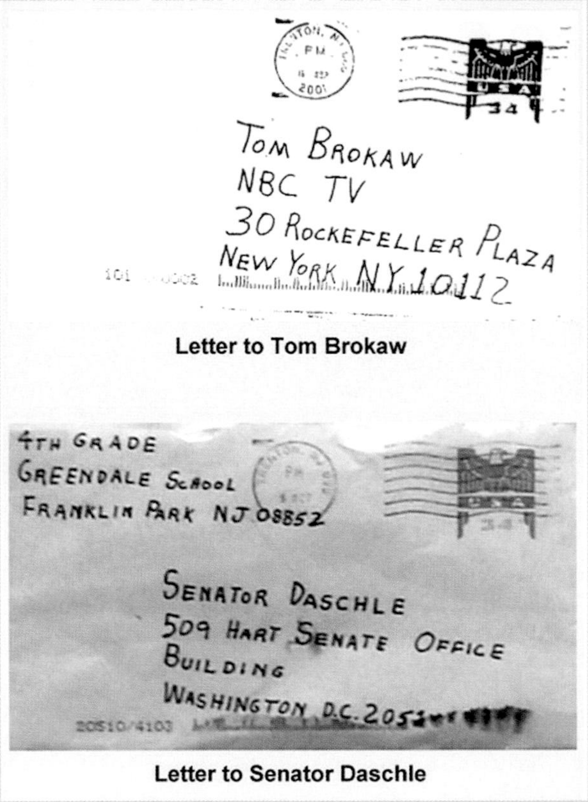

Letter to Tom Brokaw

Letter to Senator Daschle

The FBI released these pictures of anthrax-laced letters sent to Senator Tom Daschle and to NBC New York. The letter to Brokaw read: "THIS IS NEXT . . . TAKE PENACILIN [SIC] NOW . . . DEATH TO AMERICA . . . DEATH TO ISRAEL . . . ALLAH IS GREAT."

U.S. and six labs overseas. Radiocarbon dating established that the anthrax was cultured no more than two years before the mailings.

Mohammed Atta and some of the other September 11 terrorists were known to have explored using poisons from a crop-dusting plane in Florida. Other evidence pointed to some of the 9/11 hijackers having suffered from anthrax infections that needed powerful antibiotics.

Nevertheless, based on case studies, handwriting and linguistic analysis, forensic data and other evidence, FBI authorities publicly concluded after only five weeks of investigation that the person behind the recent anthrax attacks was a lone wolf within the United States who had no links to terrorist groups but was an opportunist using the September 11 hijackings to vent his rage. They did not attribute the attacks to Al Qaeda, Saddam Hussein, or any other foreign terrorist organization.

Examination of the letters apparently failed to reveal any fingerprints, smudges, hair, DNA (from licks on the stamps or envelopes), telltale paper marks, or other normal forensic evidence, indicating that the sender likely took some precautions. However, the specific strain of anthrax appears to have borne earmarks tracing it to USAMRIID at Fort Detrick. The letters contained about seven to ten grams of material, of which roughly two to three grams in the later letters were highly weaponized spores. The spore powder also contained chemical additives developed specifically for weaponization. All of the letters that were identified had been mailed between September 18 and October 9, 2001. The investigative techniques used included handwriting analysis, linguistic analysis, toxicology, criminal behavior profiling, fingerprinting, trace analysis, timeline analysis, geographic profiling, epidemiology, and many other scientific methods.

Attorney General John Ashcroft went so far as to identify publicly an individual by name as a "person of interest," but he was never charged, and the agents also raided the home of another scientist without filing any charges against him either.

Some have seen something far more complex and potentially sinister than merely a failure by so many official bodies to solve a case. Stephen P. Dresch, ex-member of the Michigan House of Representatives, argued in December 2001 that in effect the FBI et al were looking in the wrong place, and the lone wolf explanation didn't work.

Dresch highlighted in particular the "failure to consider those engaged in the development and use of lethal biological agents." He pointed out somewhat obliquely the privatization—before the attacks—of the State of Michigan's Biological Products Laboratory, whose principal function was the production of anthrax vaccine for the Department of Defense. His theory is one of many unproven and probably unprovable conjectures from journalists, scientists, and amateur sleuths, which include Israeli intelligence as the sender, a clandestine British bioweapons program as the source, and, unsurprisingly—just as the letters themselves claim – a follow-up to 9/11 by individuals assocated with Al Qaeda.

The anthrax terror case prompted one of the most far-reaching forensic investigations in history. But at the time of writing, the mystery remains unsolved.

Trace Evidence

Although it is often invisible to the naked eye, tiny bits of hair, fibers, glass, paint, explosive residue, dirt, or other types of trace evidence can link the crime scene to the criminal as neatly as a handwritten signature—provided it can be properly collected, preserved, and analyzed.

Without skilled, well-trained and equipped criminalists and lab technicians, such little stuff would never amount to much. But when it is painstakingly used to its full potential, trace evidence is big.

Retrieving a piece of glass from the clothing of a suspect in a case of assault. Can this piece of glass be traced back to the scene? Even if the fragment cannot be linked simply from its shape to, say, a bottle used in the attack, then it is possible that its type can be analyzed to the same end, using the Refractive Index (RI) temperature variation method (see page 187 for an explanation).

Trace Evidence Analysis: Waco Inferno, 1993

The controversial police assault on a compound of religious zealots in Waco takes the lives of 75 civilians, a third of them children, and leaves many wondering who was at fault.

On Sunday morning, February 28, 1993, agents of the federal Bureau of Alcohol, Tobacco, and Firearms (ATF) attempted to serve an arrest warrant for Vernon Howell, a/k/a David Koresh, and a search warrant at the Branch Davidian compound near Waco, Texas. The arrest warrant charged Koresh with unlawfully possessing fully automatic machine guns and a destructive device.

As heavily armed agents in full protective gear approached the remote compound, they came under heavy gunfire and four ATF agents were killed and fifteen wounded. The incident set off a massive siege of the Mount Carmel Center that served as the headquarters of the Branch Davidians, a religious sect that was founded on a strong belief in the Second Coming of Christ and the battle of Armageddon. Over the next 51 days no fewer than 719 law enforcement personnel from the FBI, ATF, Texas Rangers, and other agencies surrounded the embattled site.

The Branch Davidian's Mount Carmel compound near Waco is engulfed in flames, April 19, 1993. One vital question for investigators was how did the fire start? Were there FBI "pyrotechnic devices" at the three sources of the fire, as rumored?

ATF members attend to a comrade shot during the siege; the shocked faces, the use of a duvet to keep the injured man warm, these things point to a situation that was out of control. The bloody siege began with the killing of four ATF agents.

Trained negotiators conducted hundreds of conversations with some of the sect's leaders, including Koresh and his second-in-command. Members of the news media were held back out of firing range.

Although 44 members left the compound and surrendered to police, dozens more—many of them parents and children—remained inside the fortified structure. Media coverage from the site increasingly reported agents' concerns about a core of doomsday fanatics who were heavily armed and committed to violent prophesies. Rumors circulated about Koresh practicing child abuse and polygamy with his followers.

Then, at dawn on April 19, federal authorities mounted what appeared to be the beginning of an armed attack, using combat engineering vehicles and other military tactics. Soon major fires broke out inside the Davidians' stronghold. A raging inferno quickly consumed the entire complex, burning everyone inside. The remains of 75 persons were eventually recovered and removed for autopsy to determine their identity and how they had died.

In the wake of the deadly attack, which many critics claimed could have been averted, federal authorities were blamed for failure to provide fire protection or medical services, and intense controversy arose over how the fire had started. A massive crime scene investigation commenced, conducted by the Texas Rangers with assistance from the FBI and other agencies.

The Rangers divided the physical area into sectors, rows and grids, then formed teams comprised of Rangers, FBI, and other technicians, and other law enforcement agents. The teams systematically combed through each sector, identifying each item they found and pinpointing its location by sector, row, and grid number. Each item (or group of items found at a particular point) was assigned exhibit numbers and photographed. At the conclusion of each search a crime scene report was prepared listing all the items found in the search of that particular team's sector.

By May 3 the Rangers had recovered 305 firearms and approximately 1.9 million rounds of "cooked off" or spent ammunition from the compound. Some of the firearms included 20 fully automatic AK-47 assault rifles; 12 fully automatic AR-15 assault rifles; two .50 caliber semi-automatic rifles; and antitank armor-piercing ammunition.

The crime scene search also turned up thousands of items, including hundreds of exploded shells, fired shells, and bullets; Kevlar helmets and vests; camouflage outfits; hand grenades; pistols; rifles; shotguns; rocket projectiles; gas masks; chemical warfare suits; military assault knives; and fuel cans. Inside the concrete storage bunker, searchers found a large concentration of weapons and ammunition—along with 32 bodies. Many of the victims had died of suffocation. One report estimated that 390,960 rounds of ammunition were discovered inside.

The Tarrant County Medical Examiner's office, assisted by a team of anthropologists from the Smithsonian Institution, assisted in recovering the remains of the persons killed during the fire, as well as the remains of those Davidians killed during the February 28 shootout whose bodies had been buried just outside the compound. The remains were taken to the Medical Examiner's office, where autopsies and identifications were conducted.

The Medical Examiner concluded that 75 persons died inside the compound during the fire. Forty could not be identified, but positive identifications were made for 32 adults and three children. Koresh was one of those identified from dental records. His body was found in the communication room on the first floor of the building, near the door, with a bullet wound in his forehead. A rifle barrel was found on the floor near his body. Autopsies were conducted for all of the deceased. Toxicological reports noted that carbon monoxide was found in 50 of the bodies, at saturation levels varying from 10 to 79 percent. The remains were also checked for the presence of other substances, such as drugs, explosives, and gasoline.

One of the forensic anthropologists who was brought in to help identify the dead, Emily Craig, later wrote a book about her experience in which she described picking up Koresh's skull to examine the hole in the forehead—an opening that was beveled inward and surrounded by soot, showing where the bullet had entered after being fired from close range. And at the lower back part of the skull near the spinal cord she saw the exit wound. "No FBI agent could ever have gotten close enough to press a gun to his skull," she wrote.

The Rangers assembled a team of independent arson investigators to examine the cause of the fire. The team consisted of Paul Gray (Houston Fire Department); William Cass (Los Angeles Fire Department); John Ricketts (San Francisco Fire Department); and Thomas Hitching

(Alleghany County, Pennsylvania Fire Department). The team also used a specially trained chemical accelerant detection dog (and two dog handlers) from the Alleghany County Fire Department. A Texas Ranger Sergeant assisted the team.

The team based its conclusions on their examination of the scene, the dog alerts to various items of evidence found at the scene and to various items of clothing worn by survivors of the fire, and videotapes of the fire provided by the FBI, including an infrared aerial video.

Based on this evidence, the arson team found that the fire was deliberately set by one or more persons inside the compound. The fire had three separate points of origin. The arson investigation established that those fires occurred in areas significantly distant from one another, but within such a short time frame that it was not possible for the fire to have been accidentally set or for it to have been caused by a single ignition. Given this short lapse of time, and the distance between the three separate points of origin, the arson team concluded that the fire could not have been caused by accident. The blaze spread very rapidly, fed by high winds.

An arson detection dog alerted the team to the presence of chemical accelerants at numerous points throughout the compound, including at the three points of origin. The dog was also exposed to various items of clothing taken from the survivors of the fire, and the dog alerted to the presence of chemical accelerants on several pieces of that clothing.

Finally, the arson team addressed whether the fire could have been started by the FBI's deployment of tear gas into the compound. The team concluded that "the fire was not caused by nor was it intensified by any chemicals present in the tear-gassing operations."

The release of several lengthy government reports failed to quell mounting criticism of the FBI and ATF, however. Many conspiracy theorists complained that the fact that the ATF had bulldozed the crime scene on May 12, only three weeks after the attack, meant that it was trying to cover up something.

The controversy continued, fueled six years later by accusations leveled by the FBI's former deputy assistant director, Danny I. Coulson, who alleged that agents at the scene had fired two devices known as M651 CS tear gas grenades from FBI grenade launchers hours before the compound erupted in flames, thereby contradicting the FBI's official story.

The *Dallas Morning News* later reported that it had consulted a small-arms and ammunition expert with Jane's Defense Information who examined a crime scene photograph showing a device found at the compound. He noted its distinctive design—a two-toned, gray-and-gun-metal canister ringed with a bright red band—was unique to U.S. military pyrotechnic tear gas grenades. "The color coding is indicative of a 40 mm CS grenade," said the expert, Charles Cutshaw. Other reports surfaced that the FBI Laboratory had detected traces of the gas, and covered it up. But the FBI continued to deny the allegations.

Exactly two years after the tragedy, a truck full of explosives leveled the Alfred P. Murrah Federal Building in Oklahoma City. The perpetrator was a disgruntled former Army soldier who had pledged to get back at the government for what it had done at Waco (see page 62.)

Explosives: World Trade Center Bombing, 1993

The bombing of the World Trade Center in 1993 killed six persons and injured 1,000, but investigators feared it signified something far more deadly.

On Sunday, February 28, 1993, an intrepid ten-man team of forensics investigators said goodbye to their comrades and entered the black void beneath the embattled World Trade Center. Two days after the bombing, the structure was still unsound and blanketed by darkness, making it the most dangerous crime scene any of them had ever encountered.

Heading first for level B-2 near the driving ramp to B-1 directly below the cavernous ballroom of the New York Vista Hotel, the searchers struggled under the heavy weight of protective suits,

FBI and ATF agents descend into the still dangerous Twin Towers parking garage. The explosion of a 1,500-pound urea-nitrate bomb had created a crater several stories deep and 200 by 100 feet wide. No forensic evidence would definitively place those convicted of the crime at the scene; but what the team did find in the crater would lead inexorably to them.

cameras, lights, evidence kits, and communications gear. They had to be careful not to upset any tottering steel beams and concrete slabs or to step into the abyss, all the while hoping to avoid any potential undetonated explosive or invisible deadly chemicals or biological agents. Parts of the area were awash in raw sewage from ruptured pipes and toxic particles hung in the air.

The huge bomb that had killed six persons and injured hundreds of others had left a trail of crumpled cars, broken concrete, and jagged steel at its periphery. The center was a crater seven stories deep that had been gouged into the foundation of the 110-stories-high Twin Towers looming above. The blast had ripped a hole two hundred feet across the bowels of the structure, leaving much of its inner workings gutted and exposed, with steel rods and wires hanging like torn intestines. Bits of glass and shards of metal crunched under their boots as they proceeded closer and closer to the spot where the bomb was detonated.

The column of elite agents from the FBI, the Treasury Department's Bureau of Alcohol, Tobacco, and Firearms, and the NYPD's fabled bomb squad crawled on their hands and knees to scoop up samples of debris that hopefully would help them to determine more about the makings of the weapon that had tried but so far failed to bring down America's most famous colossus and symbol of western world economic might. Their primary mission was to take swabbings of explosive residue in an effort to pinpoint the blast site.

Debris from the February 26 explosion accumulates outside the World Trade Center. Workmen gradually cleared the rubble through a large hole opened over the site. The FBI and ATF agents worked round-the-clock to glean evidence before the trail went cold.

The agents worked the site in stages, moving as quickly as possible, racing against the clock to determine the extent of the damage and to unearth clues that would identify the enemy who had tried to carry off one of the worst mass murders in history. As many as 50,000 persons or more could have been killed by the towers' collapse. It was a miracle only a handful had perished. At the same time, the team had to be careful not to overlook anything and meticulously to collect and record all of their evidence.

The bomb attack was an act of

unbelievable horror, a crime against humanity that aimed to kill as many innocent civilians as possible. And the culprits were still on the loose.

By Monday the evidence team had filled nearly 40 32-gallon plastic bags with carefully selected and catalogued objects, and taken countless photographs and readings of the temperature and air contents. On Tuesday they supervised the removal of several damaged vehicles from the blast area—a spot where many of them had themselves parked their own cars in the past, a garage that suddenly had been turned into an inferno. Bit by bit, they moved through the wreckage.

Everyone scoured the vicinity for a piece of numbered metal or a scrap of paper that would lead them to the truck or explosive ingredients that caused all of the damage. Every vehicle identification number (VIN) had to be checked, every numbered auto part examined.

Bomb expert Joe Hanlin of the ATF had worked on explosives assignments all over the world for more than 20 years. But he'd never seen such devastation. Hanlin bent down and examined a twisted piece of metal he had caught in a light sweep. "This is something we need to take," he told the others. It turned out he was right.

Two other agents, John Goetz and David Sherman, found other pieces of the same metal. From the blast signature on the frame rail, experts back in the NYPD crime lab determined that they were parts of the vehicle that had carried the bomb.

After explosives experts gathered more components of the vehicle that contained the explosive device, and a painstaking search gathered enough of the truck's VIN to commence a search, investigators immediately focused on a Ford Econoline E-350 Van, bearing an Alabama license plate number of XA 70668. Their finding cracked the case.

The van data was traced to Ryder Truck Rental Co., who pointed the search to an office located in Jersey City, New Jersey, where an individual named Mohammed A. Salameh had signed a one-week rental agreement on February 23, 1993.

Salameh had later reported the van stolen from a grocery store lot on February 25 and requested a refund of his $400 cash deposit. Agents scurried to man the rental office in the event he returned.

To their surprise, on March 4, 1993, Salameh showed up at Ryder to claim his refund, and they promptly arrested him. A quick search of his residence turned up wiring, electromagnetic devices, and other suspicious material.

Salameh proved to be a 25-year-old Palestinian with a Jordanian passport. At an address in Brooklyn he had listed on his auto license, police arrested another man, Ibraham A. Elgabrowny, for attacking an FBI agent during a search of the apartment.

Elgabrowny turned out to be a relative of El Sayyid A. Nosair, another Middle Easterner who was serving a prison sentence in New York for the 1990 assassination of Rabbi Meier Kahane, the New York leader of the militant Jewish Defense League. The FBI's search of Elbabrowny's premises uncovered five fake Nicaraguan passports for Nosair and other clues.

Mohammed Salameh, one of the four men convicted for WTC 1993 bombing. He returned to collect the deposit on the van he had rented that was used to carry the bomb.

Further investigation also quickly led to the Al Salam Mosque in Jersey City that was the home base of Sheik Omar Abdel Rahman, a blind Muslim cleric from Egypt who had been tried and acquitted of involvement in the 1981 assassination of President Anwar el-Sadat and suspected in other radical Muslim attacks. Jersey City was only four miles away from the Trade Center and it appeared to have served as a gathering place for Islamic radicals.

Back at the crime scene, evidence teams remained in the crater for 26 days, picking up other traces. But the most critical piece of evidence, the VIN of the Ford van, was what had put police on the right trail.

Bomb specialists searched for traces of common high explosives, such as EDNG, NG, TNT, PETN, RDX, and others.

Experts eventually determined that the 1,500-pound urea-nitrate bomb was carried into the Trade Center garage in the rented Ryder van and parked illegally in a carefully chosen location. The device consisted of four cardboard boxes were packed into the back of the van, each containing a mixture of paper bags, old newspapers, urea and nitric acid. Next to them were placed three four-foot-long red metal cylinders of compressed hydrogen, and four large containers of nitro-glycerine, with Atlas Rockmaster blasting caps connected to each. Somebody had lit the fuses and fled before the explosion—it did not appear to have been a suicide bombing. Upon detonation, the device caused extensive damage and injuries but failed to bring down one tower onto the other. The towers had survived—at least, for the time being.

Elsewhere, police found tanks of hydrogen that two of the suspects had kept in a Jersey City storage locker.

A letter denouncing U.S. policy in the Middle East, signed by the "LIBERATION ARMY FIFTH BATTALION," that was received by *The New York Times* four days after the World Trade Center attack, warned of "more than 150 suicidal soldiers ready to go ahead" if the U.S. government didn't heed its demands. This letter was traced via computer to one of the suspects taken into custody after the bombing.

The Justice Department proceeded with a successful prosecution. On May 24, 1994, during the sentencing of four of the convicted World Trade Center bombers, Federal District Judge

Kevin T. Duffy commented that the bombers had incorporated sodium cyanide into the bomb with the intent of killing thousands of people in the towers. But this information was mostly overlooked in the media coverage.

At first, the alleged mastermind of the plot remained at large. Ramzi Yousef, a naturalized Pakastani citizen who had entered the U.S. in 1992 on an Iraqi passport and attended a training camp in Afghanistan, was identified as the brains behind the operation and the one who had taken the bomb-laden van into the garage. Immediately after the explosion, he had fled to Pakistan where he later attempted to assassinate the then-Prime Minister. Agents learned Yousef was a master terrorist with at least 12 known identities and connections to several shadowy organizations including one that came to be known as Al-Qaeda.

After a worldwide manhunt, in 1995 Yousef was captured abroad and returned to the United States to stand trial. He too was convicted and is now serving a life term in federal prison.

The prompt discoveries by forensic teams in the wake of the bombing amounted to one of the finest examples of scientific detective work in twentieth-century history. It also uncovered a network of international terrorists who appeared able and determined to strike again.

The bomb materials had cost about $300. Fortunately, the sodium cyanide added, designed to diffuse through the ventilator and elevator shafts, burned up in the explosion. This—and the fact that Yousef could not afford a bigger bomb— limited the carnage.

DNA and Dental Identification: The September 11, 2001, Attack

Combing the ruins of the former World Trade Center for bodies and evidence poses history's greatest crime scene challenge.

On September 11, 2001, American Airlines flight 11 departed Boston for Los Angeles, was hijacked by suspects armed with knives, and crashed into the south tower of the World Trade Center in lower Manhattan. United Airlines flight 175 departed Boston for Los Angeles, was hijacked and crashed into the north tower of World Trade Center. American Airlines flight 77 departed Washington Dulles for Los Angeles, was hijacked and crashed into the Pentagon outside Washington. United Airlines flight 93 departed Newark for San Francisco, was hijacked and crashed into a field in Shanksville, Pennsylvania.

In stunning succession, within the space of a couple of hours, two of the tallest buildings in the world were reduced to rubble and the nation's military and political headquarters had suffered a disastrous surprise attack. The shocking events left thousands of innocent victims dead or missing, many more thousands injured, and billions of dollars worth of property damage. The whole nation was reeling. As much of the drama unfolded on television, along with voice recordings from inside the hijacked planes and video of several of the suspects taken as they had boarded the flights, it quickly became clear that the attacks had been carried out by 19 highly trained terrorists from Al-Qaeda, who had turned commercial airliners into mighty missiles and killed themselves in the process.

When Ramzi Yousef was flown back to the U.S. to be tried and convicted of "seditious conspiracy" to bomb the towers in 1993, he remarked as he flew over them that next time, they would bring them both down.

In the disaster's wake, the crime scenes of the most spectacular coordinated terrorist attacks in history were primarily treated as urgent rescue and recovery missions.

At the same time, within hours of the attack, Attorney General John Ashcroft announced that a massive federal criminal investigation was under way and authorities had established crime scenes in New York, Washington D.C., Pittsburgh, Boston, and Newark.

The full resources of the Department of Justice, including the Federal Bureau of Investigation, the Immigration and Naturalization Service, the U.S. Attorneys offices, the U.S. Marshals Service, the Bureau of Prisons, the Drug Enforcement Administration, and the Office of Justice Programs, were deployed to investigate the crimes and assist victim survivors and victim families. Teams from the federal Environment Protection Agency and military personnel were also dispatched to the sites.

At the World Trade Center, detectives from the New York Police Department's Crime Scene Unit entered the scene literally moments after the initial attack. In the disaster's wake, at least 50 investigators worked around the clock for months documenting mountains of evidence. Other state and local police also assisted.

In Washington D.C., a multi-disciplinary team of more than 50 forensic specialists, scientists and support personnel from the Armed Forces Institute of Pathology (AFIP) swung into action in response to the Pentagon attack. Code-named "Operation Noble Eagle," the team consisted of forensic pathologists, odontologists, a forensic anthropologist, DNA experts, investigators, and support personnel, who worked for over two weeks at the Dover Air Force Base Port Mortuary at Dover, Delaware, to identify the 188 victims of the attack.

When remains arrived at the AFIP morgue, technicians scanned them for unexploded ordnance or metallic foreign bodies. FBI experts collected trace evidence to search for chemicals from explosive devices and conducted fingerprint identifications. Forensic dentistry experts from

A different skyline not only for all New Yorkers but also for the world. Plans for the 16-acre site are still not finalized, though a memorial plaza should open in 2009. The sheer scale of the forensic task was mind-boggling.

United Airlines Flight 175 impacts the South Tower. This was not a crime scene that could be isolated in the conventional way. The primary impetus was rescue, then recovery of bodies. Conspiracy theories (some quite insane), needless to say, are legion.

the Department of Oral and Maxillofacial Pathology performed dental charting and compared the data with antemortem dental records. Other personnel conducted full-body radiographs to aid in the identification process. Forensic pathologists autopsied the dead to determine the cause and manner of each death. When necessary a forensic anthropologist tried to determine the race, sex, and stature of victims who could not be identified. Forensic photographers documented injuries and personal effects. Tissue samples were collected for DNA identification and further toxicologic studies. And mortuary specialists embalmed, dressed, and casketed the remains prior to releasing them to their next of kin.

Case Study: The September 11, 2001, Attack

Among the human remains retrieved from the Pentagon and Pennsylvania crash sites of September 11 were those of nine of the hijackers, but without reference samples from the hijackers' personal effects or from their immediate families to compare with the recovered DNA, these remains could not be matched to individuals.

All 40 victims in the Pennsylvania crash and all but five of the 184 victims at the Pentagon site were identified.

The drastic nature of the World Trade disaster, however, made identification of many victims and all of the hijackers virtually impossible.

The handling of the Trade Center disaster/crime scene involved the use of a massive dump where all materials and debris from the site were taken for examination and possible disposal. The Fresh Kills Landfill on Staten Island, located about five miles from ground zero, received huge cargoes of rubble carted in by truck or barge. The millions of tons of material included everything from melted steel beams and concrete slabs to human body parts, personal effects, and other detritus of mass destruction—charred fire engines and flattened police cruisers, fragments of fire fighters' equipment, and airplane parts that were swept from the mountainous ruins created by the towers' collapse.

This recovered material created a gigantic documentation challenge for the Crime Scene Unit. At the start, as many as 300 detectives toiled with the aid of specially constructed conveyer belts, sifting machines, and other equipment designed to help them comb the crime scene for evidence and biological remains: bits of teeth, clumps of human tissue, severed fingers or charred skulls. Each possible body piece had to be identified as human, photographed, labeled, logged, and refrigerated for transport to the city morgue.

Advanced forensic DNA identification was used as much as possible to help identify many of the victims. The small size and degraded nature of much of the specimens recovered from the towers site required analysis using mitochondrial DNA using capillary electrophoresis (CE), a technique that enabled technicians quickly and economically to identify DNA in bone, blood, semen, saliva, and hair. Findings from remains recovered from the site were compared to samples obtained from hair combs, razor blades, and other sources provided by next of kin. DNA was extracted at the New York State

NOAA (National Oceanic and Atmospheric Administration) map of the site, integrating GPS and LIDAR (light detection and ranging) laser terrain modeling. This indicated the volume of debris and location of stairwells.

Police Laboratory in Albany. For comparative purposes, swabs were also taken from the mouths of consenting closest relatives and submitted to the lab.

Many of the efforts to identify remains of the Trade Center victims were spearheaded by a private genetic testing company, Myriad Genetics Inc., headquartered 2,000 miles away in Salt Lake City, Utah, because it was capable of conducting automated laboratory analysis faster than any government agency. (Similar identifications for the victims of the Pentagon attack were handled by Celera Genomics of Rockville, Maryland.)

After a year of work, the Fresh Kills depository had received 1.62 million tons of material, including 19,000 body parts, which helped to identify 1,215 victims.

The challenge was so great that it wasn't until 2005 that a final official tally was made: 2,749 persons had died in New York—147 of them as passengers or crew on the two flights, 412 as rescue workers who rushed to the scene to help, and the rest as workers or visitors who were trapped in the huge complex. Most were never identified.

Solving the crime was another matter. Investigators found evidence to support the conclusion already reached by millions who had witnessed the explosions on TV—that the explosions at the

Two days after the attack on the WTC, a wall at Bellevue Hospital is covered with posters of the missing. Mostly through DNA, dental records, and incredibly painstaking work over many months, about 57 percent of the victims were positively identified. Over 10,000 samples that could not be identified will remain at Ground Zero, available for further examination if DNA analysis techniques improve.

Trade Center resulted from the impact of two heavily fuel-laden airline passenger jets that were flown into the towers at a high speed, like guided missiles. The South Tower (II WTC) began to collapse 56 minutes 10 seconds after being struck by United Airlines. Flight 175. The North Tower (I WTC) began to collapse 102 minutes 5 seconds after being struck by American Airlines Flight 11.

Crime scene investigators normally employ photography to help document a crime scene, recording evidence as they start from the outside of the scene and then work inward to the core area. Numerous photographs are then taken from distant, mid-range, and close-up perspectives.

Ground Zero, however, was no ordinary crime scene, and many traditional rules could not be applied from the beginning. After the towers collapsed, rescue and recovery efforts assumed paramount importance.

Once the area was secured, crime scene teams entered the site and attempted to begin photographing it as best they could, although the area remained extremely hazardous. As evidence was located and removed, it was taken to another location for more thorough documentation using more sophisticated equipment. The NYPD Crime Scene Unit used four Polaroid Macro 5 SLR instant cameras that were equipped with five built-in lenses that enabled the photographers to capture details not detected by the naked eye, such as initials engraved on jewelry or a piece of a hip replacement with a serial number on it. Many of these images proved useful in making identifications.

Meanwhile, officials in New York also banned all unauthorized photography of the crime scene and set about confiscating any unauthorized cameras and film they encountered.

Much of the police photographers' work required the keeping of fastidious logs and reference numbers indicating the location where every item was found, and date and time it was found. Large sequential files were compiled to enable researchers to conduct rapid searches of the images.

Elsewhere, the identities and movements of the hijackers were tracked. For more than a year, Ground Zero remained roped off to outsiders and memorialized as the world's most famous crime scene.

Port Authority identification cards retrieved from the Fresh Kills excavation site on Staten Island. On July 15, 2002, the task of sifting through the 1.62 million tons of debris officially came to an end.

Matching

Forensic scientists sometimes have claimed they were able to positively link a suspect to the crime scene, by comparisons of blood, fibers, hair, wood, tool marks, bite marks, or other evidence, only to have a defense lawyer effectively attack their conclusions.

The attack can be based on error rates, a lack of empirical support about the reliability of the technology, or other factors. Today's courts have become more skeptical about such "matching" claims, and many expert witnesses now try to avoid asserting without equivocation in their testimony that they have made a match, or that there is no possibility for difference or error.

Sandia National Laboratories investigated natural reflectance of organic matter—fingerprints, blood, semen—in the 1990s. The major government-owned/contractor-operated (GOCO) organization has been involved in projects as diverse as thermonuclear weaponry, microelectromechanical technology, and a decontamination foam to neutralize anthrax in buldings on Capitol Hill in 2001.

Trace Evidence: Wayne Williams

Police said they had solved the "Atlanta Child Murders" case by matching fibers and hairs with mathematical precision.

In the summer of 1979 police discovered the decomposing bodies of two missing African-American boys, aged thirteen and fourteen, in a wooded area of Atlanta, Georgia. Over the next two-and-a-half years, as many as 29 more black bodies, mostly male children, turned up around Atlanta, triggering a racially charged scare that quickly became a national nightmare. Most of the victims had been strangled. The motive was unclear. Many feared a wave of terror by the Ku Klux Klan, but there were no firm leads.

With over 25 Atlanta children listed as dead or missing, possibly at the hands of a serial killer, police finally caught a break. On May 22, 1981, a patrolman staking out a section of the Chattahoochie River heard a splash directly underneath the nearby Jackson Parkway Bridge.

The officer responded to find the only car on the bridge at that time was being driven by a young man, Wayne Bertram Williams, 23, a local black freelance photographer and self-styled music promoter. Williams was stopped and quizzed, but released. His behavior seemed suspicious: he'd said he was trying to locate a young female musician, but the information he gave as her telephone number and address proved non-existent. When asked what he had thrown in the river, Williams had replied, "Just trash." Police then dragged and searched the river in that area, but found nothing.

Two days later, however, the naked body of 27-year-old Nathaniel Cater was pulled from the Chattahoochie, a mile downstream from the bridge. Cater had been reported missing for a few days. The coroner ruled his death a homicide by asphyxiation. In his hair investigators found a single strand of yellowish-green nylon fiber with a tri-lobed cross section.

Police began to link similar fibers to some of the other murders. Williams was put under surveillance

Wayne Bertram Williams giving a press conference at the Fulton County gaol, Georgia, December 13, 1983. Convicted of two murders and blamed for at least 23; but was there more than one killer on the loose in Atlanta?

and police obtained a search warrant. They combed his car and house and removed thousands of fibers from the home's olive-colored carpet. Working in the Georgia State Crime Laboratory, Detective Larry Peterson peered through his microscope to discover that hairs on some of the victims matched hairs that had been taken from Williams' German shepherd and violet acetate bedspreads in Williams' home. Hal Deadman, from the FBI's Microscopic Analysis Unit, also examined the fibers and hairs and agreed with Peterson about the matches. After identifying the carpet manufacturer as West Point Pepperell Corporation of Dalton, Georgia, the detectives began to investigate how common the "Luxaire" line was. The company said only 16,500 square feet of the carpet in that color had been made, between December 1970 and December 1971. Police computed the odds of another house in Atlanta having the fibers to be 7,792 to one.

Also, other fibers found on some of the bodies were matched to carpets in Williams' car. Police estimated such a match would have occurred in fewer than one in 3,828 cars in the city. In all, some of the bodies had ten different fibers that matched fibers found in Williams' home or car. The chance of someone else in Atlanta having the same fibers in their home and car carpets was estimated at 29 million to one. And so, on June 21, based almost exclusively on the fiber and hair evidence (but also the admission of testimony linking him to earlier victims), Williams was arrested for Cater's slaying. Williams protested his innocence, but his coworkers claimed to have seen him with some of the victims and others reported having noticed scratches on his forearms at various times—possibly a clue from some of his strangling victims. Although Williams was suspected in most or all of the child murders, he was tried only for Cater's killing and the homicide of 21-year-old, Jimmy Ray Payne, whose body was found in the Chattahoochie not far from the site where Cater was discovered. On mostly circumstantial evidence, he was convicted of both killings on February 27, 1982, and sentenced to two consecutive life terms.

The Wayne Williams case proved to be a classic trace evidence case. Trace evidence includes materials that are transferred from one person and place to another, the presence of which may be detected only through microscopic analysis. Such an approach was developed by Edmond Locard (1877–1966), who posited that "every contact leaves a trace." In many ways, Locard's "exchange principle" stands at the heart of forensic science.

LINGERING DOUBTS

More than twenty years later, controversy continues to surround the Atlanta Child Murders case, particularly those for which no one was ever convicted. Williams has always maintained his innocence. Although the Georgia Supreme Court upheld his conviction in 1984 and rejected an appeal for a new trial in 2001, he is still pursuing appeals in the U.S. District Court in Atlanta.

Some victims' parents refuse to believe that Williams was responsible for any of the killings, while others contend he was not the only murderer. Former FBI profiler John Douglas has said he doubts that Williams committed all of the slayings, but he believes evidence points to Williams in at least eleven of the child killings. Chet Dettlinger, a former assistant to the Atlanta Chief of

Police who later wrote a book about the case, contends that glaring errors were made in the fiber matching analysis: fibers taken from the house or car ignore other persons who may have also shared those environments, or introduced some of the matching hairs or fibers from other places.

In May 2005, DeKalb County police Chief Louis Graham, who has investigated the episode for many years, reopened five of the "Atlanta Child Murder" cases, saying he never believed that Wayne Williams had committed any of the killings. A five-member cold case squad is assigned to review the matter. Both prosecutors and defense attorneys say nobody yet knows what forensic evidence for the case still exists and what its condition is, especially blood and physical specimens.

Williams' appeals attorney, Michael Lee Jackson of Buffalo, has argued that if DeKalb police can subject the alleged dog hair evidence to DNA analysis that did not exist during Williams' trial, they could destroy that pattern of evidence. "The pattern was based on incredibly bad science, and fallacious extrapolation of evidence," Jackson said. "These fibers were like a fingerprint. If you show on some of these victims that the fingerprints were actually somebody else, then that destroys the entire pattern." The original investigators were limited to microscopic analysis; new scientific methods of blood analysis and other specialties could come up with more precise results.

Dekalb's first black sheriff, Sidney Dorsey, was a partner of Graham's in the Atlantic homicide squad as far back as the 1960s, and says he entered the Williams' home with a search warrant for the telltale carpet. Back at the lab, he says Larry Peterson later told him: "Those fibers are flying everywhere." He says Peterson implied the fibers were nowhere near as rare as they were made out. What can be made of this pretty damning story? In light of the fact that Dorsey was given a life sentence in 2002 for ordering the murder of the man who beat him for sheriff, not much.

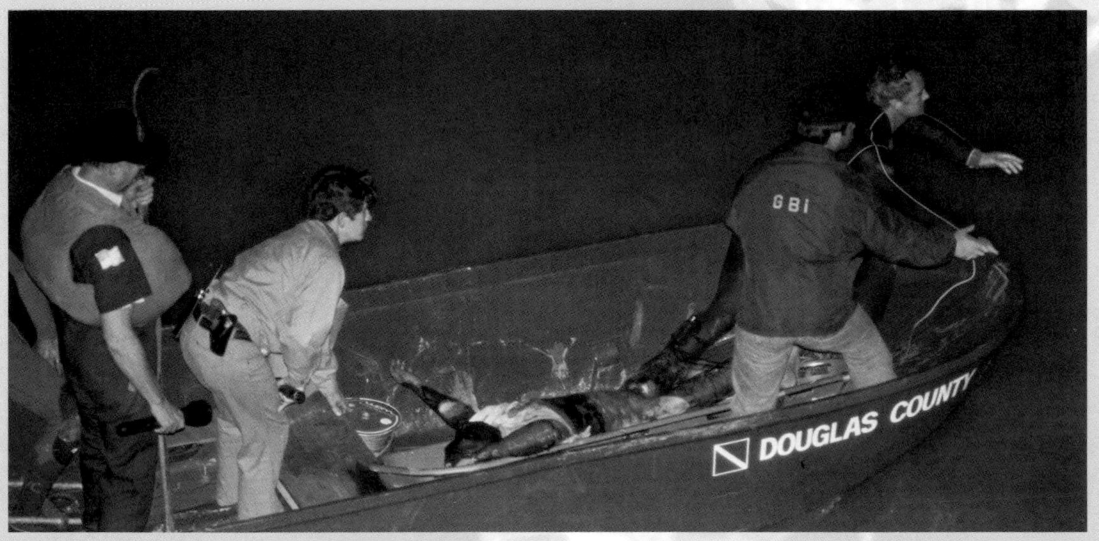

A victim is recovered by police. Apart from the fiber evidence, it was the decision of Judge Clarence Cooper to alllow pattern evidence to be given against Williams that convicted him; never charged with a child murder, the jury heard of his links to child victims.

Teeth Marks

One of the oldest, yet sometimes still controversial forensic techniques involves teeth mark analysis. Just how reliable is this different kind of dental record? Teeth have been used for centuries to identify human remains. Paul Revere, the Revolutionary-War Boston patriot and silversmith, once made a set of ivory and silver dentures for a friend, Dr. Joseph Warren. After the British killed Warren at the Battle of Bunker Hill in 1775 and threw his corpse into a mass grave, Revere helped identify Warren by his dental work so he could be reburied in the family plot. To this day, dental records often prove invaluable for identifying individuals in mass disasters, war, and other causes of death.

In 1849 dental evidence was first accepted in a United States court after a Harvard professor, J.W. Webster, turned up missing and some charred bone fragments were found in a suspect's laboratory. Investigators later matched them with the victim's dental records and the defendant was convicted of murder.

By the late 1930s some police detectives were using teeth to do more than identify human remains. These pioneers began to view teeth like other tools that could leave a distinctive mark—something that could assist them in solving certain crimes, particularly rape and sex murders, battered-child homicides, and cases in which the victim left a defensive bite mark on the assailant. Usually the front teeth, top and bottom, were found to leave the strongest impressions.

Police also began to utilize other tooth marks they had found at crime scenes, such as a piece of chewing gum bearing telltale impressions that had been left next to a cracked safe, or a half-eaten chunk of cheese at the scene of a restaurant burglary. Crime scene photographers took pictures of such evidence in an effort to record their unique or distinctive features. Sometimes they compelled a suspect to bite plastelina in order to make a plaster cast of the markings, or they consulted dentists to take advantage of the latest dental cast-making technique.

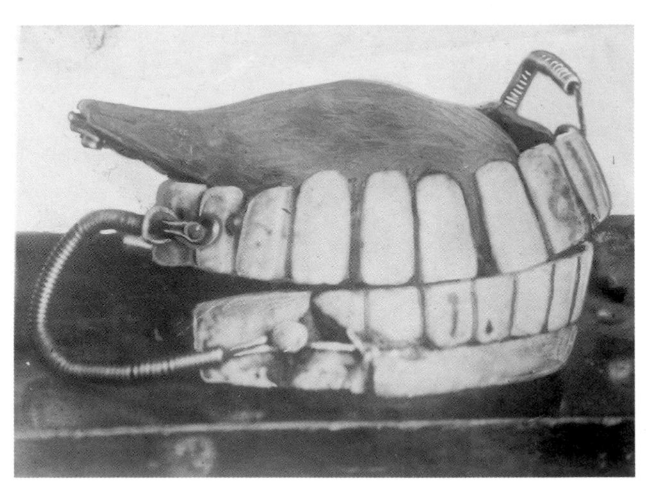

One of the moat convenient identification details a body can furnish: false teeth. If he hadn't died in bed of the croup, with his wife and Dr. Craik at his side, these would have identified George Washington.

By the 1930s some prosecutors began to present dental evidence in court in the form of enlarged photographs of the casts in which the characteristic points were noted in red ink along with precise measurements, just as they did with fingerprints. In the late 1940s and early 1950s, noted UK forensic pathologist Keith Simpson wrote a number of influential articles asserting that bite-mark evidence could prove as accurate as fingerprints in matching crime scene evidence to a defendant. In 1967 a fifteen-year-old schoolgirl, Linda Peacock, was murdered in Biggar, near Edinburgh, Scotland. Odontologist Dr. Warren Harvey took dental impressions of 30 men from a nearby detention center. With Simpson's help, a bite mark on the body was linked to one man at the center, 17-year-old Gordon Hay. One of his teeth was pitted because of a disorder known as hypocalcination and the pitting fit the impression on the girl's body exactly. The defense attempted to have the evidence ruled inadmissible but failed, and bite-mark testimony could now be introduced in subsequent cases.

Forensic odontologist Dr. Lowell Levine photographs the jaw and teeth of one of the Romanovs in 1992, more than seven decades after the Russian royal family were executed during the Revolution. The historic identification was positive.

Experts soon agreed that human bite marks left on foodstuff, such as cheese or gum, could offer a three-dimensional impression, which was superior to the two-dimensional impression made in human tissue, although a landmark case in California involved a criminal who left a clear three-dimensional impression in a victim's nose.

Forensic scientists learned that some bites might penetrate a person's skin, but leave only bruising, while the blood marks of a bruise could be mistaken for the impression of a tooth. They also discovered that bite marks could also change over time, particularly when human tissue was involved. As a result, the interpretation of bite marks came to be viewed as an art-based science.

Dentists pointed out that restorations, fillings, rotations, tooth loss, breakage, and injury can make one person's teeth unlike anyone else's. Yet attributing a bite mark to any one tooth can prove difficult.

Over the years, some homicide investigators came to believe that bite mark evidence was usually confined to certain types of crimes and offenders. Some tried to interpret bite marks in an effort to gain insights into a criminal's mind-set. They thought bite marks in homosexual

Mob boss Stefano Magaddino allegedly had Albert Agueci killed in 1961. The body was found on a farm in Rochester, NY. Thirty pounds of flesh were missing from the body, which had been doused in gas and burned. Agueci was identified by his few remaining teeth. Nothing else could have.

homicides, for example, often tended to be found on the victim's back, arms, shoulders, face, and scrotum. Breast and thigh bite marks, they believed, might indicate heterosexual aggression and tend to have been made more slowly and sadistically, which may have left a better impression. Battered-child victims, it was said, might exhibit randomly placed bite marks that were generally diffuse and of poor detail.

Although Texas began allowing bite-mark evidence in 1954, prior to the 1970s, bite-mark evidence wasn't treated very seriously in some courts. In 1970 a few dental specialists created their own division of forensic odontology in the American Academy of Forensic Sciences, and some of these practitioners began to communicate with each other nationally on various cases. Then U.S. appellate courts began to allow bite-mark evidence in certain instances on a case-by-case basis, generally if they involved three-dimensional patterns in foodstuffs or wounds.

From the mid-70s, a handful of newly established "forensic odontologists" helped to decide a spate of high-profile cases including Ted Bundy in Florida, Lemuel Smith in upstate New York, Francine Elveson in the Bronx, Karla Brown in Wood River, Illinois (whose body was exhumed but still revealed telltale impressions that led to a conviction of the murderer), and Jeffrey Dahmer, the Milwaukee serial killer and cannibal. Seventeen of Dahmer's victims were identified through dental records, despite his use of muriatic acid to remove flesh from his victims' bones.

By the early 1990s developments such as computer image enhancements, developments in color photography, and the introduction of an L-shaped ruler for use in comparing marks, were thought to have greatly enhanced the quality of forensic odontology, so that even a state such as New Jersey, which had remained skeptical, in 1993 finally accepted bite-mark evidence as a method of proving an assailant's identity. By 2000 there were about 500 professionals in the United States doing forensic dental work, including 100 who were certified by the Board of Forensic Odontologists; although some critics complained about weak standards for certification, noting that only two bite-mark cases were required.

Then disaster struck. DNA testing proved some of the bite-mark experts had been wrong. Bite evidence had been used to send Ray Krone to death row in Arizona for the 1991 murder of a

Phoenix cocktail waitress, but DNA later cleared him and he was released after 10 years of incarceration. Kennedy Brewer had been condemned to death in Mississippi based on bite marks, but DNA exonerated him. Dale Morris Jr. of Florida was arrested in 1997 based on a bite mark that two forensic dentists used to connect him to the rape, torture, and murder of a nine-year-old girl, but he was later set free by DNA.

In the wake of such acknowledged errors, some observers began to question the scientific basis of bite-mark "matching." One leading critic, Dr. C. Michael Bowers, an odontologist and lawyer who served on the examination and credentialing committee of the American Board of Forensic Odontology, in 2002 co-authored a study that estimated the performance of board-certified odontologists in a workshop exercise. Bowers figured that on average, they falsely identified an innocent person as the biter nearly two-thirds of the time. At the end of the twentieth century, the use of bite-mark analysis was touted as a powerful new tool to take a bite out of crime. But in the face of renewed scrutiny, its role in forensic odontology was diminished.

At the same time, the longstanding use of teeth to identify individuals became faster and less cumbersome. Since the crash of TWA Flight 800 off Long Island in 1996, digital dental radiography has increasingly been used to identify human remains in mass disasters. New digital X-ray systems and portable X-ray tube heads allow forensic odontology teams to record victims' teeth directly at the site, without the need for a light screen, and then transmit the images via secure satellite to other sites throughout the world, in order to instantaneously match dental records. The dental identification unit at Dover Air Force Base used digital radiography on September 11, 2001, to identify some victims from the flights that crashed in Pennsylvania and Washington, and this technology is now an integral part of the portable morgue used by the Federal Emergency Management Agency's Disaster Mortuary Response.

Necrophiliac and cannibal Jeoffrey Dahmer habitually cooked his victims and used acid to dissolve parts of the bodies. so it's not surprising that dental records were the most important identification method for his seventeen victims. He killed for the first time when aged just eighteen.

Teeth Marks: Lemuel Smith

How Lemuel Smith's teeth marks condemned him not once, but twice.

As the police tried to solve the gruesome murder of a Schenectady woman, Marilee Wilson, in 1976, a New York State Police investigator was struck by what appeared to be distinctive bite marks on her body—the signature of a psychotic killer.

Later, when police identified a suspect, Lemuel Smith, they convinced a judge to order Smith to give bite impressions, and the results helped to convince them to charge the ex-convict with murder. At Smith's trial, the state called to the stand an expert witness, Dr. Lowell Levine, a forensic odontologist, who testified that the bite marks on the victim matched the teeth of the

Forensic odontologist Dr. Lowell Levine's testimony helped to convict serial killer Ted Bundy for the murder of Lisa Levy in Florida. When asked to provide a dental impression after his arrest, Bundy refused. A search warrant allowed the Pensacola investigators to pursue the match. Dr. Richard Souviron, a dentist from Coral Gables, took the picture of Bundy's teeth above.

defendant. After hearing the testimony, Smith confessed that an inner demon, "John" had carried out the crime, and Smith was convicted and sentenced to prison.

The story of Smith's teeth might have ended there except that in 1981 the battered corpse of a missing correction officer, Donna Payant, from Greenhaven Correctional Facility in Dutchess County, New York, was discovered in the prison's landfill. Examination of the body revealed that she had been strangled and sexually mutilated: her nipples had been bitten off and she had other bite marks on her flesh. But there didn't appear to be any blood or other physical trace evidence to solve the murder.

When investigators and a newspaper reporter discovered that Smith was one of the inmates in Greenhaven prison, and that he had been in the same area as the victim at the time of her disappearance, attention immediately focused on the apparent human bite marks. Speculation grew that the crime seemed to carry Smith's modus operandi.

Smith's lawyer, William Kunstler, tried to block the bite evidence from being introduced, but the trial judge ruled it admissible. Once again, Dr. Levine took the stand to testify as an expert witness against Smith.

> "Did there come a time when you made a scientific comparison between the bite mark on the Marilee Wilson case and the bite mark that you found on the Donna Payant photography?" the prosecutor asked him.
>
> "Yes sir," Dr. Levine replied. "It is my opinion that the two bite marks were made by the same set of teeth."

In his judgment, Smith's bite had left similar impressions on the bodies of both victims.

Prosecutors later called another witness, Dr. Neal Riesner, to the stand. Although originally hired by the defense as a bite expert, Riesner revealed in the courtroom that he had examined the photographs and concluded with a reasonable degree of medical certainty that the marks matched Lemuel Smith.

Kunstler presented his own alternative theory that others, including prison staff, had committed the Payant murder and conspired to pin the blame on Smith. But largely as a result of the bite evidence, Smith was convicted and sentenced to death, although the sentence was later struck down and converted to life in prison. He has remained in solitary confinement ever since. Dr. Levine was later appointed codirector of the New York State Police Forensic Unit and became generally regarded as one of the world's leading forensic odontologists. His work has spanned the globe.

Forensic Laboratories

The proliferation and enhancement of forensic labs throughout the country has transformed crime fighting. One of their most scientifically complex challenges is in the discovery and identification of toxins in the deceased.

Sir Arthur Conan Doyle, the English physician who wrote the Sherlock Holmes mysteries in the late nineteenth and early twentieth century, is generally credited with conceiving the notion of a modern scientific laboratory to aid in solving crimes. Twenty-three years after Doyle's introduction of Holmes, the world's first crime lab was established in France, in 1910.

In 1924 the Los Angeles Police opened the first police crime laboratory in the United States. Northwestern University School of Law followed in 1929 by creating the first independent crime lab, the "The Scientific Crime Detection Laboratory."

Several large cities and states, as well as the FBI, started their own facilities as well. The New York State Police, for example, formed a facility in 1936, staffed initially by two full-time chemists and eight assistants, clerks and typists who were hired through the federal Works Progress Administration.

These personnel were supplemented by various scientists and technicians hired on a per diem basis. Much of their early work focused on research and data collection in ballistics, photography, hair sampling, blood grouping, and other disciplines. The staff painstakingly examined and catalogued headlight lenses, cloth fibers, paints, tire tread patterns, and other diverse items for use in solving hit-and-run accidents and other crimes.

A typical laboratory at the start of the twentieth century. The pipettes, flasks and fumigation cupboard would play a part in forensics later; but what the scientist primarily needed was a dedicated space.

In 1942 the Illinois State Police established a new milestone by creating the first mobile crime lab. It contained an X-ray unit, photography, fingerprinting and polygraph equipment, microscopes and supplies that allowed scientists to conduct nitrate and blood stain tests, identify fluids, restore serial numbers erased from metals, and conduct qualitative and quantitative analysis of unknown substances.

As late as the 1960s, however, many of the forensic labs contained little more than a fingerprint kit, a polygraph machine, a Breathalyzer, some magnifying glasses, a microscope, and a kitchen sink. But then spending on crime fighting began to dramatically increase. By the mid-1970s, 47 states had crime labs. Amid today's forensics craze, many major crime labs teem with mass spectrometers, infrared spectroscopes, neutron activation devices, X-ray machines, elaborate gas tanks, and other state-of-the-art technology. The largest facilities are associated with states, cities, and large police agencies, but in recent years the number of moderate-sized labs has grown exponentially to include many that specialize in drug testing, serology, or DNA, and more of these labs are under way.

One of the joys of the Holmes novels is of course their forensics exactitude. Fact follows fiction, in that the real dedicated forensics lab antedates its description by Conan Doyle.

By 2002 the federal government conducted the first national census of publicly funded crime labs and found there were 351 federal, state or local forensic labs scattered throughout the country. These included 33 federal, 203 state or regional, 65 county and 50 municipal labs with about 9,400 full-time employees.

A typical forensic laboratory started the year with a backlog of about 390 requests, received 4,900 requests, and completed 4,600 requests. Forty-one percent of publicly funded labs in 2002 reported outsourcing one or more types of forensic services to private labs—particularly DNA-related casework requests.

The labs varied widely in their capacity to conduct various types of analyses. About nine in 10 labs could identify controlled substances, whereas about six in 10 could conduct biology screening, firearms and toolmarks analyses, crime scene evidence collection, latent print analysis, or trace evidence assessments (such as paint chips or other non-biological materials). About half the labs could process DNA evidence and conduct toxicology analyses. About one in four labs reported being able to examine questioned documents and about one in nine said it had the capability to conduct forensic computer analyses.

Nationwide, crime laboratories in 2002 received about 2.7 million requests for forensic laboratory services and they were able to process just under 2.5 million of those requests during the year. The most frequently requested forensic laboratory service by far, the identification of controlled substances, resulted in nearly 1.3 million requests during the year or about half of all requests. Toxicology samples (468,000) and latent print requests (274,000) were the next most common types of samples for which laboratory analyses were requested. Law enforcement agencies submitted about 61,000 requests for DNA analysis—about 2 percent of all requests for laboratory services, to publicly operated crime labs—just under 42,000 of these were processed during the year.

About nine in 10 labs that handle fingerprint identifications reported having the capability to process fingerprints utilizing the automated fingerprint identification system (AFIS) that checks an electronic fingerprint database.

Sixty-one percent of the nation's labs in 2002 were accredited by the American Society of Crime Laboratory Directors Lab Accreditation Board, and an additional 10 percent were accredited by some other organization. Most labs reported that, consistent with standards in the field, they adhered to established standardized protocols for fingerprint examination, controlled substance analyses, and DNA testing.

The public's image of scientific crime solving is embodied in the modern forensic laboratory—a sprawling expanse of work stations full of white-coated scientists who are envisioned hunched over microscopes and computers, holding test tubes to the light, examining slide trays, applying chemicals to tagged clothing, test-firing guns on the indoor range, and carrying out myriad other complex tasks in a constant search for the truth.

The first forensics lab (arguably) was organized by Dr. Edmond Locard in France in 1910, inspired by Sherlock Holmes. It didn't take much organizing, just a microscope and a spectroscope.

It is there that scientists detect the signature of a serial bomb-maker, or discover the telltale ink that reveals a forgery, or match a weapon found in a suspect's car to a cartridge found at the murder scene.

In fact, much of the work performed at such forensic labs is clinical, tedious, and routine. Yet it is increasingly important in battling certain kinds of crime, such as illegal drugs, sex offenses, homicide, and now of course, terrorism.

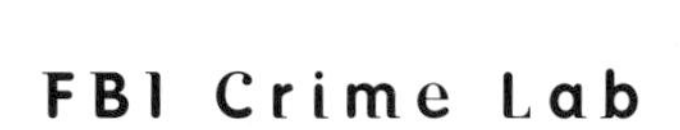

FBI Crime Lab

The FBI Laboratory has grown to become a comprehensive, state-of-the-art forensic resource for law enforcement throughout the U.S. It has not been without its critics, however, even from within.

Since its beginning in 1932 the Federal Bureau of Investigation Laboratory has become the world's premier forensic science

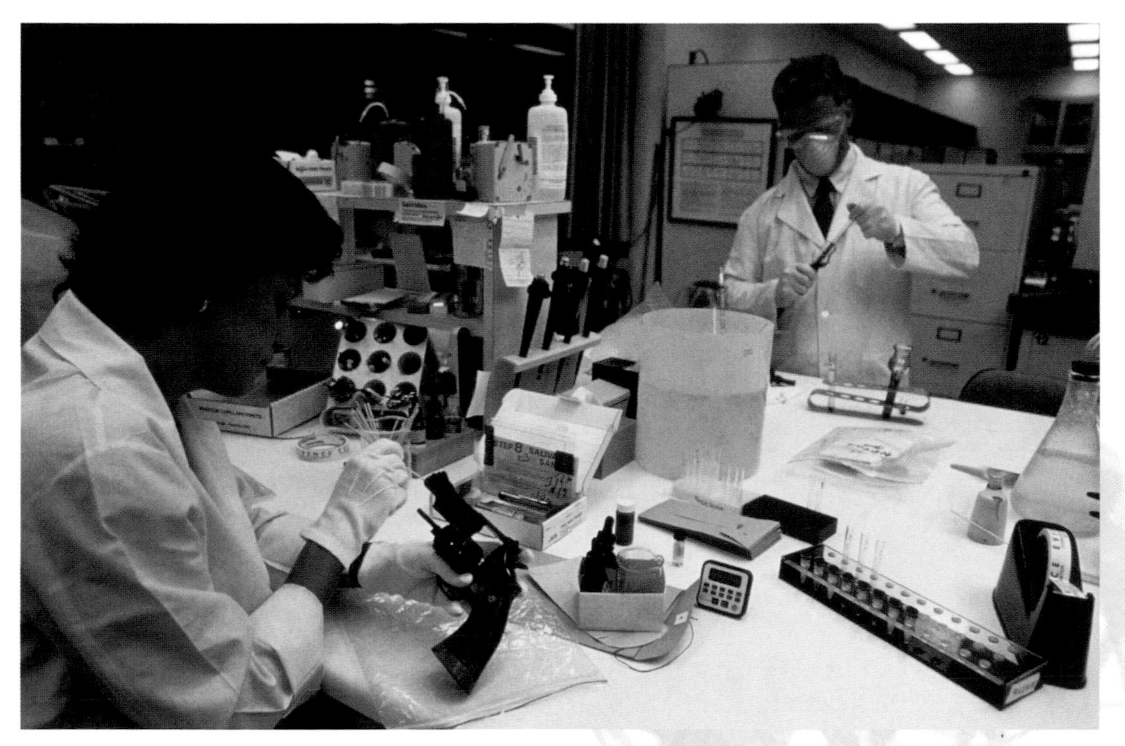

Forensic scientists at work at the FBI serology lab, Washington D.C. At left, the .357 Magnum is being examined for traces of blood. At right, the technician is dissolving a blood clot to be analyzed for blood group or DNA sequencing.

headquarters, performing more than a million annual examinations in a wide range of fields, including: chemistry; computer analysis and response; DNA analysis; explosives; firearms and toolmarks; forensic audio, video, and image analysis; hazardous materials response; latent prints; materials analysis; questioned documents; special photographic analysis; structural design; and trace evidence. The FBI lab is actually a collection of related state-of-the-art specialized laboratories and facilities, including:

- CODIS—The Combined DNA Index System is a program that facilitates the electronic sharing of information by outside state and local labs. This system provides forensic labs with software that enables them to access databases of convicted offenders, missing persons, and unsolved crimes. DNA profiles may be exchanged and compared between labs that are trying to link suspects to crime scenes.
- NDIS—The National DNA Index System is part of the CODIS system and allows DNA profiles from convicted offenders to be accessible to forensic labs.
- IAFIS—The Integrated Automated Fingerprint Identification System allows latent fingerprint comparisons to be made between labs. IAFIS is the largest database of its kind.

The FBI Lab also provides advanced education, training, and technical support to state and local labs, and encourages the development of the forensic science field.

In recent years, however, the lab's reputation suffered some setbacks. During the Oklahoma City Bombing prosecution, an FBI scientist, Frederick Whitehurst, publicly alleged FBI laboratory specialists and technicians had fabricated or suppressed evidence, given perjured evidence and obstructed justice in thousands of cases to favor the prosecution. His whistle blowing raised a stink.

Some of Whitehurst's concerns were reinforced in a 500-page report issued in April 1997 by the Inspector-General for the Department of Justice, who found the lab was responsible for errors in testimony and substandard analytical work, but didn't find FBI crime lab agents had committed perjury or fabricated evidence. The Inspector-General's inquiry primarily concerned only three units of the Laboratory—the Explosives (EU), Chemistry-Toxicology (CTU), and Materials Analysis (MAU) units.

Some of Whitehurst's allegations concerned the O.J. Simpson case. To address the defense's contention that the police had planted blood at the crime scene and on socks found in the defendant's residence, prosecutors in the case asked the FBI Laboratory to determine whether the blood preservative EDTA was present in those blood stains. CTU Chief Roger Martz and several research chemists at the FBI Forensic Science Research Unit (FSRU) at Quantico, Virginia, worked to develop a method for identifying EDTA in blood. Whitehurst alleged that scientists at the FSRU had commented that Martz had committed perjury, misled the jury concerning the validation studies conducted by the FSRU scientists, misled the defense by stating that all digital data from the analysis of the evidence had been erased, and generally testified in an arrogant manner.

The Inspector General found no basis to conclude that Martz committed perjury or any corroboration that FSRU scientists had made such allegations. Nor did he find that Martz improperly erased digital data. But the IG did criticize Martz for his deficient record-keeping and note-taking and for the manner in which he testified, which "ill served the FBI because it conveyed a lack of preparation, an inadequate level of training in toxicological issues, and deficient knowledge about other scientific matters that should be within the expertise of a chief of a unit handling chemical and toxicological analyses in the Laboratory."

Some of Whitehurst's most serious allegations involved the Oklahoma City Bombing case—principally, that David Williams (the examiner from the Explosives Unit who had offered conclusions about the type of explosive used in the attack), had based some of his conclusions on speculation from the evidence associated with the defendants. The IG agreed with Whitehurst and censured Williams and his supervisor.

A series of disclosures documented other instances of malpractice, flawed science, doctored lab reports, posed evidence, shoddy investigative work, false testimony, and cover-ups involving the FBI Lab, some of them involving higher-ups within the FBI.

Toxicology: Dr. Carl Coppolino

Forensic toxicology cracks the perfect crime. One of the greatest cases in forensic toxicology occurred in the 1960s, involving Dr. Carl Coppolino, an anesthesiologist who was suspected of murdering his lover's husband and—in a different romantic entanglement—his own wife, without leaving any trace.

The investigation started when Marjorie Farber, a widow in Sarasota, Florida, told police that her former lover, Dr. Carl Coppolino, had secretly murdered his wife in Florida. When Mrs. Farber was challenged, she added, "And what's more, two and a half years ago, he murdered my husband. I know he did it, because I was present when he did it."

The accuser was first pegged as a jilted lover with an axe to grind. But after a check revealed that her late husband, Col. Bill Farber, a successful insurance executive, had indeed died in New Jersey of a "heart attack"—the same cause of death as Dr. Carmela Coppolino—the Sarasota authorities contacted Dr. Milton Helpern, the New York City Medical Examiner and one of the world's top experts in forensic pathology, for assistance. Helpern took on the challenge.

Dr. Carl Coppolino on his way to court for the first time, December 9, 1966, accused of murdering retired army Lieutenant Colonel William E. Farber. He would be acquited on that charge, but found guilty of murder in the second degree of his own wife Carmela.

According to Farber, after her husband's murder, the Coppolinos had moved to Florida and she had followed them by relocating to a nearby town. Carl had indeed asked his wife Carmela for a divorce, but not so he could be with Farber. Instead, he had fallen in love with a rich divorcee, Mary Gibson. Carmela had refused and soon afterward she died in her sleep. Five weeks later, Carl married Gibson.

Helpern and his associates were highly suspicious, but both deaths had been officially certified as being from "natural causes," and in order to prove any murder charges against Coppolino, they would need hard evidence.

Because Carmela's body had never been autopsied, the authorities decided first to exhume her remains. The death had occurred in Florida and the disinterment took place in New Jersey, but the autopsy was conducted in Manhattan. Helpern's autopsy revealed an apparent needle puncture mark in the left buttock, a healthy heart, and no discernible cause of death. A later autopsy on Farber's husband produced evidence of death by strangulation, which was consistent with her story that Coppolino smothered him in his sleep. As a result, grand juries in New Jersey and Florida indicted Coppolino for homicide.

Coppolino hired the flamboyant lawyer, F. Lee Bailey, considered one of the nation's top criminal defense attorneys. Bailey immediately attacked Mrs. Farber's motives and her credibility. His defense was so effective that the New Jersey trial resulted in Coppolino's acquittal for Colonel Farber's murder.

In Florida, however, the toxicological evidence proved decisive. The prosecution's theory was that Coppolino had injected his victims with fatal amounts of succinylcholine, an artificial curare-like drug that physicians used in light doses to relax skeletal muscles, but which in larger quantities could shut down the heart. Investigation revealed that Carl Coppolino had previously ordered a number of vials of the drug, claiming he needed it for cat experiments, and that appeared suspicious to police. But the central problem was that succinylcholine had never before been detected in the human body because it quickly broke down into succinic acid and choline, compounds that were normally present in dead tissue; they are there in such small quantities that ordinary techniques fail to detect them.

Helpern's chief toxicologist, Joseph Umberger, worked on the tissues for a long time in an effort to devise a new test that could detect abnormally large amounts of the two substances but not react with the minute quantities normally present. He developed the technique to prove there was an abnormally high concentration of succinic acid in the organs of Carmela Coppolino's body, but he could not show that there was an excess amount in the left buttock itself, because he could not apply the technique to fatty tissue.

To test their conclusions, Helpern enlisted the help of an associate, Dr. Valentino Mazzia, who had himself injected with a paralyzing dose of succinylcholine with a respirator and other safeguards present, and the results were recorded before he was helped to recover. Another researcher, Dr. Bert LaDu, was able to identify the drug in Carmela Coppolino's buttock tissue.

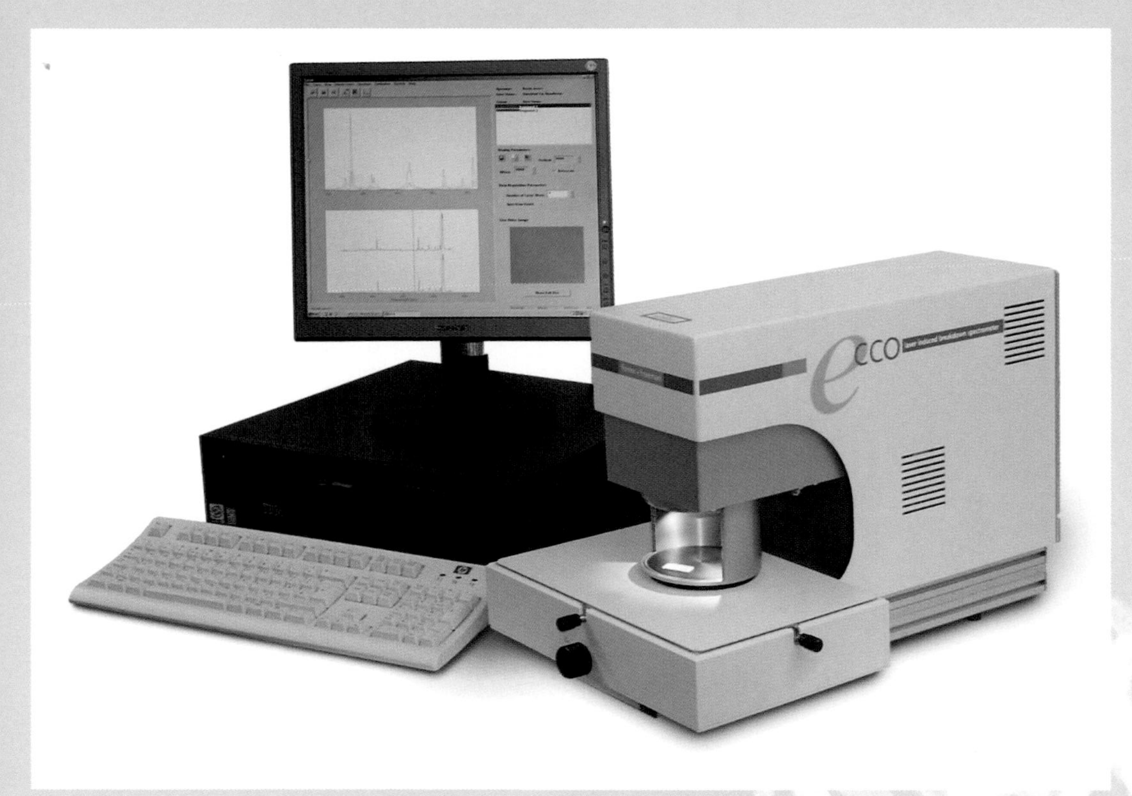

An alternative to conventional analytical techniques such as mass spectroscopy, Laser Induced Breakdown Spectroscopy (LIBS) uses a powerful burst of high intensity pulsed laser that vaporizes matter—glass, paint, or bullet fragments for example—to produce a characteristic visible spectrum as the various elements in the sample return to their natural temperature. The elements are identified and different samples matched or distinguished. The manufacturers call it the Elemental Composition Comparator (ECCO).

Based on this input, Helpern established that Carmela Coppolino had died from an overdose of succinylcholine chloride. Bailey fought every step of the way, arguing that the test was new and had not been accepted by the courts. But the judge allowed it into evidence and based on the forensic evidence the jury convicted Carl Coppolino of second-degree murder.

Forensic toxicology (the analysis of drugs and poisons in blood and body fluids) requires some of the most challenging and exacting work in forensic science. Some of the presumptive tests (screening tests) commonly used today include thin layer chromatography (TLC) and immunoassay techniques. The patented Enzyme Multiplied Immunoassay Technique (EMIT) is a fast and cheap method for screening urine and blood for drugs. Some of the common confirmatory analyses that use instrumental techniques are infrared spectroscopy (IR), gas chromatography/mass spectrometry (GC/MS), and high-pressure liquid chromatography/mass spectrometry (HPLC-MS).

DNA Revolution

Rapid recognition of the power and potential abuses of DNA has revolutionalized the field of forensic science, requiring crime scene specialists and laboratory scientists to be much more painstaking and careful in their work.

In 1984, as a sidelight to other genetic research he was conducting, Dr. Alec Jeffreys of the University of Leicester in England developed a revolutionary new identification method called "genetic fingerprinting," involving the deoxyribonucleic acid (DNA) molecule that makes up genes and carries inherited information. Word of the discovery spread quickly and Jeffreys' dramatic success in applying it to solve a murder-rape case in England (Colin Pitchfork), made international headlines and thrust the miracle of "DNA fingerprinting" into the world of forensic science. The first case involved a series of sex-murders of young children in the village of Narborough. After it was reported that Jeffreys had created a "DNA fingerprint," police announced they wanted to compare DNA in semen taken from the bodies of two of the victims to DNA in biological evidence received from possible suspects who voluntarily submitted samples of their blood. When a baker named Colin Pitchfork was detected trying to convince one of his coworkers to give a blood sample on his behalf, police investigated further and extracted a confession. They then enlisted Jeffreys to test Pitchfork's blood and indeed found his DNA profile matched the semen samples taken from his two victims. (Earlier, British police had asked Dr. Jeffreys to verify a suspect's confession to one of the two rape-murders. Dr. Jeffrey's tests proved the 17-year-old could not have been the perpetrator. So the first significant result or DNA fingerprinting was an exclusion.)

Professor Sir Alec Jeffreys, the English molular biologist who discovered genetic fingerprinting in 1984 at the Department of Genetics, University of Leicester. He is examining autoradiograms of DNA fingerprints.

Scientists had come to understand that the nucleus of every cell contains a string of coded information in the form of a long, ribbon-like polymer made up of repeated units called nucleotides. Each nucleotide consists of a molecule of sugar bound to another containing phosphorous and another containing nitrogen. A single DNA molecule contains millions of these nucleotides, arranged in pairs in long chains that are curved into a helix.

Vast amounts of genetic material are stored in these codes. Part of it is unique to each individual, except identical twins, thereby making it an important tool for identifying individuals—especially since this unique character is present in every living cell that has a nucleus, including blood, hair roots, or bone marrow, and also because only a tiny amount of DNA material is needed for analysis.

By virtue of Jeffreys' path-breaking research, which soon was being written up in all of the leading scientific journals, news magazines and stock market newsletters, the world came to learn some of the basics of the new identification technique. Some proponents claimed that the method was capable of establishing an individual's absolute uniqueness, but geneticists pointed out they were claiming only extraordinarily high probabilities, ranging from 1 in 100,000 to 1 in 100 million.

In the United States, instant adoption of some of Jeffreys' techniques led to forensic use of DNA testing on a limited basis. The first U.S. criminal conviction of a DNA case was that of *Florida v. Andrews* in 1986. Tommy Lee Andrews was convicted of rape by the Orange County Circuit Court through a blood sample/semen trace match. Large-scale DNA typing quickly followed. In the early 1990s, as DNA typing began to take more of a role in human identification, concerns and criticisms began to emerge.

Police quickly realized that by examining DNA taken from a crime scene, a forensic scientist could later compare it to a sample taken from the individual who was suspected of the crime, or it could be compared by computer to countless other DNA samples of convicted individuals that had been taken and stored in a database. For example, a sperm stain found on a victim's clothing, which was believed to contain the assailant's DNA, could be used for comparison—possibly to produce a match to the offender.

Every law enforcement department throughout the country—indeed, across the world—had unsolved cases that might be solved through such advancements in DNA technology. DNA promised possibly to hold the key to solving residential burglaries, sexual assaults, and murders. It suddenly seemed that the saliva on the stamp of a threatening letter, or the perspiration on a rapist's discarded mask, might hold the key to solving a crime.

In Austin, Texas, an investigator requested DNA testing on the phone cord used to choke a victim in an open case. The investigator realized that in the course of choking someone, so much force and friction is often applied to the rope or cord that a perpetrator's skins cells may rub off his hands and be left on the ligature. The rapist as well had been aware of the power of DNA and he had taken precautions to protect himself—he wore both a condom and rubber gloves. Yet during the struggle, the attacker was forced to use one hand to hold the victim down, leaving only one hand to pull the phone cord tight. As a result, the attacker had to grab the remaining end of the cord with his mouth, thereby depositing his saliva on the cord. Although the developed profile came from saliva rather than skin, DNA not only solved the case in Austin, but it also linked the perpetrator to a similar sexual assault in Waco.

DNA breathed new life into criminal investigations. Like fingerprinting, it offered investigators an opportunity to link independently and objectively an individual (perpetrator or victim) to a crime, but with a degree of certainty that was heretofore unmatched by other forms of evidence.

DNA has also been found to have many other forensic uses, such as to exonerate suspects or convicted persons, or to help investigators develop important leads. DNA analysis has also become extremely useful in forensic medicine and anthropology, as well as paternity testing. As a result, DNA has become the evidence of choice whenever blood or body fluids are involved.

Due in part to its tremendous evidentiary power, the emergence of DNA identification quickly prompted a drastic change in the level of legal scrutiny required for physical evidence. As demonstrated in the O.J. Simpson case in 1995, issues of "integrity" and "contamination" of DNA evidence became determinative, due to the sensitivity of this type of biological evidence. The ability to introduce DNA findings in court can be greatly affected by the evidence collection and preservation methods that were used, all through the chain of custody. DNA evidence that is not properly recognized, documented, collected, and preserved may prove to be of no value to a prosecution. In short, DNA has raised the bar and changed the way courts think about all forensic evidence.

Around the time of the Simpson verdict, some states began to establish genetic identification databases. In 1989 the Virginia General Assembly became the first U.S. legislature to require mandatory DNA samples from specific classes of offenders: sex offenders and certain violent criminals. One year later, it was all felons. Soon every state followed suit. Laws were passed to require designated offenders, initially sex offenders and murderers, to provide a sample of their DNA. A blood or hair follicle sample was taken from each prisoner and stored; a genetic profile of the offender was then stored in the database. Labs began to compare these profiles to DNA entered to the database from previous unsolved murders, rapes, and other crimes.

Almost immediately after initiating such a system, the State of New York, for example, in March 2000 announced that it had linked the DNA of an inmate serving time for robbery at Sing Sing prison, Walter Gill, to an unsolved murder from 1979. Other cold cases were also solved as more agencies matched some of their crime scene DNA to known offenders.

Leon Dundas of Florida had been killed in a drug deal in 1999. Remembering that Dundas had refused to give a blood sample in connection with a rape investigation in 1998, investigators were able to obtain the deceased Dundas' blood sample through the medical examiner's office and they forwarded it to the DNA lab at the Florida Department of Law Enforcement. Dundas' DNA profile was compared with the national forensic index and a match was made between

Dundas and DNA evidence from a rape victim in Washington D.C. The FBI then entered DNA evidence from additional unsolved rapes committed in Washington. Dundas' DNA matched seven additional rapes in Washington and also three more in Jacksonville, Florida. Police in Washington said that without DNA, they would have never identified Dundas, who had no prior recorded history of violent crime.

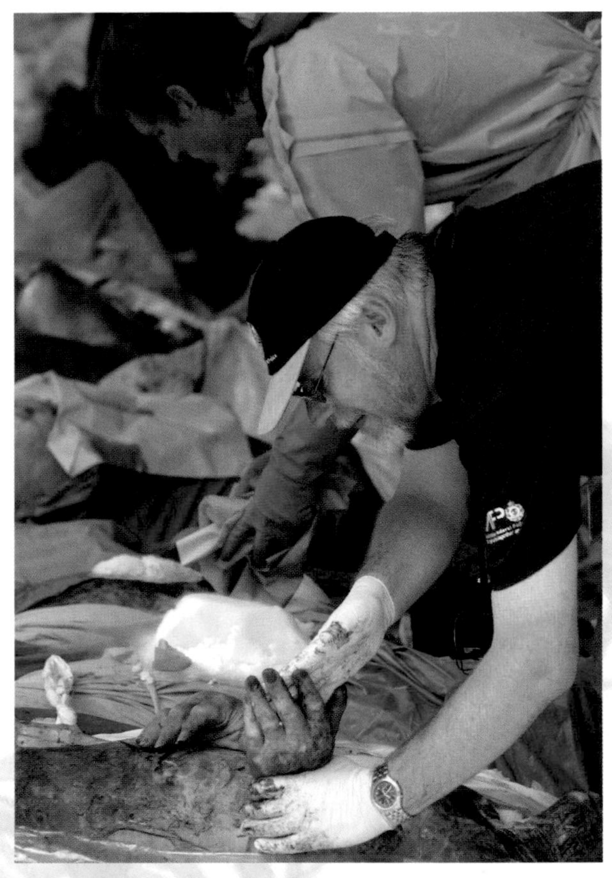

An Australian forensic scientist takes DNA samples from a victim of the terrible December 26, 2004, tsunami in Thailand. Unlike most other countries hit, where most of the dead were buried or cremated without identification, in Thailand the biggest Disaster Victim Identification (DVI) effort in history was undertaken—at some health risk to the investigators.

In another instance in April 2001, a "cold hit" was made to the perpetrator's convicted offender DNA profile in the database in North Carolina. The perpetrator had been convicted of shooting into an occupied dwelling, an offense that required inclusion in the North Carolina DNA database, and police had obtained a sample of his blood. That sample was analyzed and matched to crime scene evidence from a series of violent attacks on elderly victims that had occurred in 1990 in Goldsboro, North Carolina. The high-profile murders and rapes had been attributed to an unknown individual dubbed the "Night Stalker." When confronted with the DNA evidence, the suspect confessed to all three crimes. Mark Nelson, special agent in charge of the North Carolina State Crime Laboratory, said, "Even though these terrible crimes occurred more than 10 years ago, we never gave up hope of solving them one day."

Integrity

Shortly after DNA arrived on the scene, some scientists began to question publicly the soundness of the DNA typing procedures used by some commercial companies, and they stressed the need for strict standards for proficiency testing to ensure that DNA examiners and labs did not commit errors that would mar the method's reliability as forensic evidence. They called for quality assurance programs in the laboratories, lab accreditation, individual certification, rigorous validation studies, and immediate training of any and all personnel involved in handling DNA at crime scenes or labs.

Calls for improved quality assurance on the part of forensic laboratories has resulted in a move toward laboratory accreditation. Some states, such as New York and California, require their forensic DNA laboratories to be accredited by the American Society of Crime Laboratory Directors' Laboratory Accreditation Board (ASCLD/LAB), which audits labs to determine whether they are meeting laboratory quality assurance standards. This development helped to professionalize DNA forensic laboratories and had a beneficial effect on crime laboratory standards generally.

The pace of this movement toward greater quality assurance and accountability has not satisfied everyone. By December 1996, only 138 of nearly 400 crime labs in the country were accredited by ASCLD/LAB—and the FBI Laboratory was not among them. There was also considerable disagreement over what form proficiency testing should take and exactly how labs and their personnel should be evaluated. Some critics complained that the lack of a requirement for forensic scientists to be licensed or certified made it less likely for the field to be regarded as highly as the medical or legal professions.

Police Lieutenant Dan Nolan stands before photographs of victims of the Green River killer. Between July 1982 and March 1984, Gary Ridgway murdered at least 48 women. Advances in DNA testing linked him to four of the bodies 20 years later.

DNA: Gary Dotson

An Illinois prisoner, convicted of rape, becomes the first person in the U.S. to be exonerated by DNA comparison.

Things are often not as they seem. Until the dawn of forensic DNA information, the state's case against Gary Dotson appeared to be like thousands of other rapes involving a hapless defendant who had been brought to justice by an effective law enforcement system.

It started on the night of July 9, 1977, when a police patrol officer in the Chicago suburb of Homewood discovered a disheveled 16-year-old girl wandering along a road, crying that she had been raped. Cathleen Crowell said shortly after she left her job at Long John Silver's restaurant, three young men had accosted her and forced her into their car. One of them tore off her clothes, raped her in the backseat, and used a broken beer bottle to carve letters on her stomach.

The girl was taken to the hospital and her parents were notified. A routine rape examination retrieved semen and hairs and the emergency room physician made a sketch of the superficial scratches on her stomach. A police artist helped the girl produce a graphic description of the assailant and police guided her through a mug book. Using these mugshots and later pointing him out in a lineup, Crowell identified Gary Dotson and he was charged with the vicious crime.

Dotson protested he was innocent, but at his trial the victim identified him in open court, saying: "There's no mistaking that face." A state police forensic scientist testified that the semen on the victim's undergarment came from a type B secretor—someone with type B blood who secreted his blood antigens into his other bodily fluids—and that the defendant was a type B secretor. The prosecution also presented forensic testimony that a pubic hair removed from the victim's underwear "matched" the defendant's and was dissimilar to the victim's.

Although Dotson's public defender contested some of these claims, the trial judge overruled all of the defense objections and Dotson was convicted. The judge sentenced him to 25 to 50 years for rape and another 25 to 50 years for aggravated kidnapping, the terms to be served concurrently. The Illinois Appellate Court upheld the conviction in 1981. There it might have ended. Dotson remained behind bars. The victim resumed her life and in 1982 she married and moved to New Hampshire. But then something happened.

Three years later, the woman told her pastor she was plagued by guilt because she had fabricated a rape allegation that had sent an innocent man to prison. She said she had invented the story out of fear that her boyfriend at the time had made her pregnant, and she thought she needed a cover story if that turned out to be the case. She said she had torn her clothing and even inflicted the superficial cuts on own her stomach to try to fortify her lie.

The pastor got a lawyer to represent her. But when the lawyer contacted the Cook County State's Attorneys Office, the prosecutors refused to reopen the case, so the lawyer finally went to a Chicago TV station, which reported the story. In defense of the DA, the flag-waving *Chicago Tribune* began attacking the credibility of the woman's recantation.

Dotson's new lawyer got the original trial judge to order a hearing, but the judge quickly dismissed new exculpatory forensic evidence that basically substantiated the woman's recantation, and ordered Dotson back to prison. With public sentiment increasingly coming round to Dotson, his lawyer petitioned for executive clemency and Governor James Thompson granted a clemency hearing. The three-day hearing became an international media circus. In the end, clemency was denied, but Dotson was released on parole, and Governor Thompson (a former state prosecutor) used the occasion to praise the state's wonderful criminal justice system.

Two years later Dotson was charged with a parole violation. His newest lawyer happened to read an article in the *Newsweek* magazine of October 26, 1987, that reported the discovery of a revolutionary new scientific technique capable of linking criminal suspects to crimes through DNA. "Every cell in an individual," the article said, "including those in blood, semen, and hair roots, has the same DNA, the molecule of heredity. Since each person's DNA is unique (unless he is an identical twin), it can be used for identification with near-perfect accuracy. British geneticist Alec Jeffreys of the University of Leicester, who discovered the technique, estimates that the odds against any two unrelated people sharing the same DNA fingerprints are billions to one." The article said the technique already had been used to resolve a handful of paternity and immigration cases in Britain and lately had successfully linked a 27-year-old man, Colin Pitchfork, to the rape-murders of two teenage girls in England (see page 146).

Sensing a possible lifesaver, Dotson's lawyer immediately filed a motion asking the presiding judge of the Criminal Division of the Cook County Circuit Court to order DNA testing in the Dotson case. After a hearing on Breen's motion for DNA on January 7, 1988, the assistant state's attorney told the judge he had no objection to DNA testing. "If there's any test out there that's going to help us come to the truth, we want to pursue it," he said.

The judge effectively left the matter up to Governor Thompson. Within a few hours, the governor's office announced that Thompson not only had approved the proposal, but also, that gubernatorial staff already had contacted Alec Jeffreys in England, who had agreed to do the DNA testing. This would be a first in American legal history.

A sample of semen from the victim's underwear was sent to Dr. Jeffreys. His patented process—restriction fragment length polymorphism (RFLP)—required DNA of high molecular weight, meaning that it would not work if the genetic material had lost mass due to degradation. Unfortunately, because Dotson's sample was badly degraded, Jeffreys' test results were inconclusive. But scientists reported there was another type of DNA testing—polymerase chain reaction (PCR)—which had recently been patented by a firm in California. It had the advantage of working on degraded samples.

DNA samples were sent to Forensic Science Associates in Richmond, California. The lab performed PCR DQ alpha tests that showed that the semen on the victim's undergarments could not have come from Dotson but could have come from the victim's boyfriend. In short, the state's case was destroyed.

Finally, nearly a year later, the state accepted the test results and on August 14, 1989, Dotson's conviction was overturned, after he had spent eight years in prison for a crime he didn't commit.

Subsequent investigation revealed that the alleged victim of the crime had lied under oath and falsely accused someone of a crime that had never occurred; the state's forensic serologist had misrepresented the forensic evidence and exaggerated his own scientific credentials; prosecutors had withheld exculpatory evidence; Dotson had received ineffective assistance from his earlier defense lawyers; and the case had been riddled by a host of other abuses and problems that never would have come to light without the arrival of a new forensic "magic bullet" called DNA. DNA had opened a Pandora's box.

By 2005, DNA was credited with exonerating 100 persons awaiting execution on death rows throughout the United States. Many of them were credited to the work of two determined defense lawyers, Barry Scheck and Peter Neufeld, who had started to use DNA evidence in a 1988 murder case from the Bronx. They later founded the Innocence Project at Cardozo Law School of Yeshiva University to exonerate wrongfully convicted persons using DNA evidence.

Gary Dotson and his false accuser Cathleen Crowell Webb are interviewed in May 1985. Dotson was found to be innocent following PCR DNA tests. The damage that Webb's false accusation caused for genuine rape victims cannot be gauged.

Mitochondrial DNA: Albert DeSalvo

Using new tools to revisit the famous case of the Boston Strangler. Could this "resilient" form of DNA prove DeSalvo's innocence?

Between June 1962 and January 1964, a rash of unsolved rape-murders of respectable women in Massachusetts gained increasing notoriety in the newspapers, prompting national news media to proclaim that the "Boston Strangler" was on the loose.

Today, DNA might have helped to crack the case, but at the time there was no such tool available. Lacking any strong physical evidence, police investigators weren't convinced that one man was responsible, but their Strangler Bureau kept checking for prints, looking for clues, and running down leads, aware that all the publicity was turning up the political heat.

Then, in March 1965, a bombshell hit. Police and the media suddenly learned that the Boston Strangler had confessed, on a Dictaphone, to F. Lee Bailey, the flamboyant Boston lawyer who was now his criminal defense attorney. According to Bailey, the Strangler admitted carrying out 13 murders, not just the 11 that newspapers had attributed to him. What was more, he was already locked up on another charge.

The self-proclaimed serial killer was Albert DeSalvo, 34, a petty criminal who had confessed in November 1964 to a series of more than 400 breakins throughout New England and two rapes. While held for psychiatric observation at Bridgewater State Hospital, DeSalvo spent time with a conniving Russian mobster, George Nassar, who discussed with him some money-making schemes. Nassar was already represented by Bailey and he put Bailey in touch with DeSalvo.

After recording DeSalvo's detailed confession, Bailey took it to the State Attorney General and played it without revealing his client's name in an effort to get them to agree not to seek the death penalty in the case and possibly to allow the killer to be declared insane. DeSalvo became publicly identified as the Boston Strangler and the cases were essentially closed when he was sentenced to life imprisonment for the other crimes and not the murders. But DeSalvo later recanted his Strangler confession. In 1973 he was stabbed to death at Walpole Prison.

DeSalvo left behind a growing group of skeptics, some of them related to the last murder victim, Mary Sullivan, who doubted he was the Strangler. Some of these supporters later claimed that, hours before his death, DeSalvo had told them he was about to release new information revealing who was the real Boston Strangler.

In 2000 the controversy surrounding the Boston Strangler case took another turn when Mary Sullivan's kin allowed her body to be exhumed, 36 years after the murder. Private investigators located a semen stain on Sullivan's body, but the stain was so degraded that further tests for the presence of spermatozoa and for prostatic-specific protein p30 (an antigen found in relatively high amounts in semen) were not possible. But the material did reveal mitochondrial DNA.

Eight of Albert DeSalvo's presumed 13 victims; Mary Sullivan is bottom row, third from left. Her nephew Casey Sherman was convinced that DeSalvo did not kill her and DNA testing almost 40 years after her death seemed to support him. "I wanted to prove that DeSalvo did do it and at least have some kind of closure. Once I peeled back the onion, I realized this guy didn't do it. The only thing connected to DeSalvo was his confessions and not a shred of physical evidence."

Mitochondrial DNA is a special type of maternally inherited DNA that is a gift to anthropologists because it does not change from generation to generation, so it can be used to trace human migrations (see glossary). It also does not degrade as quickly as "ordinary" DNA.

A private forensics team also obtained blood from DeSalvo's brother Richard, who would have had identical mitochondrial DNA to his brother. Then they compared the mitochondrial DNA (mtDNA) with mtDNA from the semen stain found on Sullivan's body. According to James E. Starrs, Professor of Forensic Evidence at George Washington University, the DNA evidence found on Sullivan didn't match DeSalvo's. State officials called the private DNA tests "interesting," but not conclusive. To date law enforcement authorities haven not announced any new developments in the nagging case of the Boston Strangler.

DNA Glossary

Alleles. Alternate gene forms or variations, which are the basis of DNA testing.

Antigens. Any biological substance, such as a toxin, virus, or bacterium, that can stimulate the production of, and combine with, antibodies. Chemicals that are attached to the red blood cells to create the different blood groups. Variances in human antigens can be used to identify individuals within a population.

CODIS. Authorized in 1994, the COmbined DNA Index System is the national DNA databank used to fight violent crime. The Convicted Offender index contains DNA profiles of individuals convicted of felony sex offenses (and other violent crimes). The Forensic Index contains DNA profiles developed from crime scene evidence. All DNA profiles stored in CODIS are generated using STR (short tandem repeat) analysis. CODIS utilizes computer software to search automatically its two indexes for matching DNA profiles.

DNA. Deoxyribonucleic acid, which contains genetic material and whose shape resembles a rope ladder that has been twisted (the double helix). An individual's DNA is unique except in cases of identical twins.

DNA match. *See inclusion.*

DNA profiling. The process of testing to identify DNA patterns or types. In the forensic setting, this testing is used to indicate parentage or to exclude or include individuals as possible sources of body fluid stains (blood, saliva, semen) and other biological evidence (bones, teeth, hair). Scientists use a small number of sequences of DNA that are known to vary among individuals a great deal, and analyze those to get a high probability of a match, or non-match.

DNA typing. *See DNA profiling.*

DQ alpha (DQa). An area (locus) of DNA that is used by the forensic community to characterize DNA. Because there exist seven variations (alleles) of DNA at this locus, individuals can be categorized into one of 28 different DQ alpha types. Determination of an individual's DQ alpha type involves a Polymerase Chain Reaction-based test.

Electrophoresis. A technique by which DNA fragments are placed in a gel and separated by size in response to an electrical field.

Epithelial cells. Membranous tissue forming the covering of most internal surfaces and organs and the outer surface of the body.

Epithelial cell fraction. One of two products from a differential extraction that removes DNA from epithelial cells before analysis of sperm DNA is conducted. The other product is the sperm cell fraction.

Exclusion. A DNA test result indicating that an individual is excluded as the source of the DNA evidence. In the context of a criminal case, "exclusion" does not necessarily equate to "innocence."

Gene. A segment of a DNA molecule that is the biological unit of heredity and transmitted from parent to progeny.

Genotype. The genetic makeup of an organism, as distinguished from its physical appearance or phenotype.

Inclusion. A DNA test result indicating that an individual is not excluded as the source of the DNA evidence. In the context of a criminal case, "inclusion" does not necessarily equate to "guilt."

Inconclusive. The determination made following assessment of DNA profile results that, due to a limited amount of information present (e.g., mixture of profiles, insufficient DNA), prevents a conclusive comparison of profiles.

Marker. A gene with a known location on a chromosome and a clear-cut phenotype (physical appearance or observable properties) that is used as a point of reference when mapping another locus (physical position on a chromosome).

Mitochondrial DNA (mtDNA). A special form of DNA found in the mitochondria of cells. (The mitochondria are not in the cell nucleus, where other DNA is found; they generate fuel for the cell's functions.) It is maternally inherited and survives much longer than ordinary DNA.

Mitochondrial DNA analysis (mtDNA). This can be used to examine the DNA from samples that cannot be analyzed by RFLP or STR. Nuclear DNA must be extracted from samples for use in RFLP, PCR, and STR; however, mtDNA analysis uses DNA extracted from another cellular organelle called a mitochondrion. While older biological samples that lack nucleated cellular material, such as hair, bones, and teeth, cannot be analyzed with STR and RFLP, they can be analyzed with mtDNA. In the investigation of cases that have gone unsolved for many years, mtDNA is extremely valuable. Comparing the mtDNA profile of unidentified remains with the profile of a potential maternal relative can be an important technique in missing person investigations.

Polymerase Chain Reaction (PCR). A technique to amplify a sample of DNA to a size sufficient for analysis and identification, used in DNA profiling.

Restriction Fragment Length Polymorphism (RFLP). A technique used in the process of DNA profiling. RFLP is used to analyze the variable lengths of DNA fragments that result from digesting a DNA sample with a special kind of enzyme. This enzyme, a restriction endonuclease, cuts DNA at a specific sequence pattern know as a restriction endonuclease recognition site. The presence or absence of certain recognition sites in a DNA sample generates variable lengths of DNA fragments, which are separated using gel electrophoresis. They are then hybridized with DNA probes that bind to a complementary DNA sequence in the sample. RFLP is one of the original applications of DNA analysis to forensic investigation. With the development of newer, more efficient DNA-analysis techniques, RFLP is not used as much as it once was because it requires relatively large amounts of DNA. In addition, samples degraded by environmental factors, such as dirt or mold, do not work well with RFLP.

Secretor. A person who secretes the ABH antigens of the ABO blood group in saliva and other body fluids.

Serologist. A forensic scientist who specializes in biological fluid analysis.

STR Analysis (STR). Short tandem repeat technology is used to evaluate specific regions (loci) within nuclear DNA. Variability in STR regions can be used to distinguish one DNA profile from another. In CODIS, the FBI uses a standard set of 13 specific STR regions. The odds that two individuals will have the same 13-loci DNA profile is said to be about one in one billion.

Y-Chromosome Analysis. The Y chromosome is passed directly from father to son, so the analysis of genetic markers on the Y chromosome is especially useful for tracing relationships among males or for analyzing biological evidence involving multiple male contributors.

Digital Evidence

Notwithstanding the DNA revolution, nothing has changed crime scene investigation as much as the still-continuing digital revolution. Not only has it given birth to new criminal methods—even new crimes—it has also provided the criminalist with a new armory and demanded new forensic methodologies. At first, reports surfaced about a handful of high-tech experts pursuing mischievous young and nerdish computer intruders, known as "hackers," who seemed to be more of a nuisance than a serious threat to society, based upon their attempts to break into large defense and corporate computer systems. Soon, however, law enforcement authorities began to take note of more insidious transgressors, who needed to be aggressively prosecuted, and to realize that computer evidence was also becoming more useful in investigations of more conventional crimes as well.

Albert Einstein said, "All our lauded technological progress—our very civilization—is like the

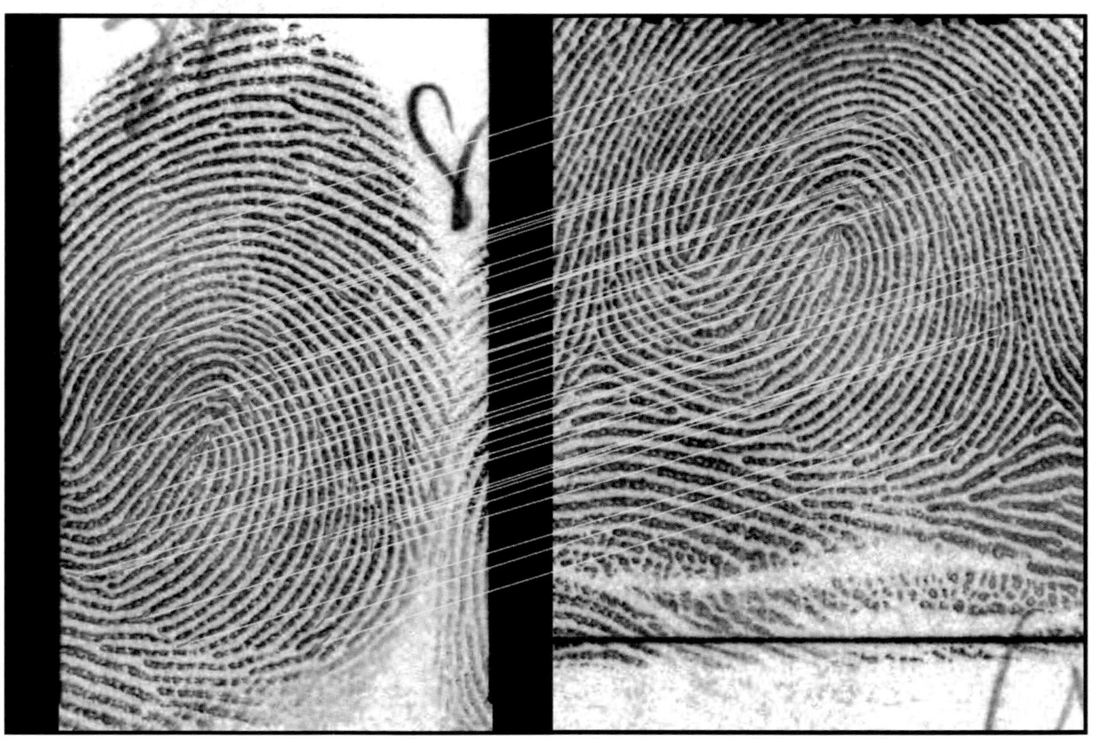

This computer generated display shows an automated match after a minutia comparison of two fingerprints. The computer found 38 minutia similarities (matches), which are identified in green. Green is also used to display the ridge flow. Yellow is used to connect the comparisons from one fingerprint to the other. A human examiner would not have found as many matches.

axe in the hand of a pathological criminal." Armed with the power of computers, some criminals already were proving themselves able to steal large sums and inflict more damage than ever before. Drug lords and spies were found to be multiplying their gains using all kinds of the latest digital technology. International terrorists were exploiting computers, electronic telecommunications devices, and the Internet to conceive and carry out attacks that posed unimaginable dangers. White-collar criminals were stealing billions at the tap of a keyboard rather than the point of a gun. Sometimes the police stumbled upon electronic footprints leading to immense criminal domains. Technicians rushed to contain computer viruses that threatened to spread like plagues, capable of destroying everything in their path unless stopped by the appropriate antiviral agent.

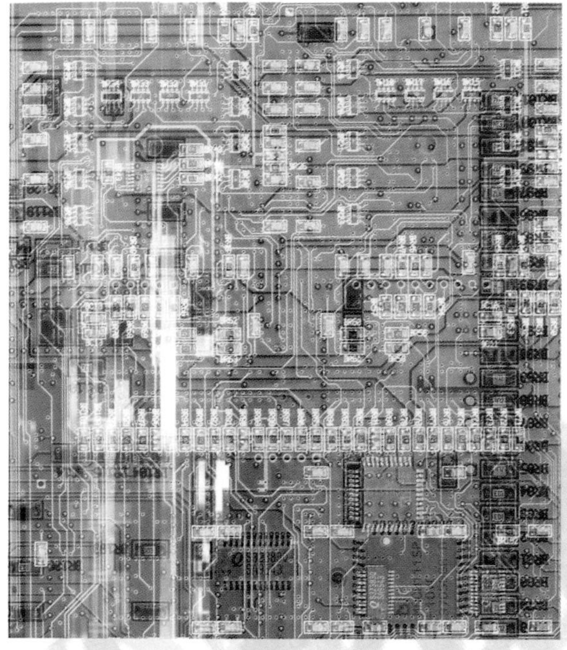

Following forensic digital tracks requires advanced computer knowledge, both of software and the hardware it runs on.

The new lexicon included concepts and terms that seemed like futuristic bursts from an Alvin Toffler best-seller: cybercrime, cyberattack, cyberstalker, cyberenemy, cyberpunk, computer crime spree, cybersurveillance, cybervandals, hacktivists, cyberterror, and cybertrail.

As police scrambled to catch up and harness computer power and expertise to catch them and other criminals, forensics specialists began to confront the dizzying array of ways in which computers and other electronic evidence could figure in crime. Computers could be the fruits of a crime, they could also provide evidence of a crime, or prove to be an instrument of a crime, a means of a crime, a target of a crime, a tool of a criminal, a repository for information related to criminal activity. Gradually, more experts came to understand that digital evidence, or evidence that is stored or transmitted using computers, can be useful in all sorts of investigations, including homicide, child exploitation, computer intrusion, and corporate malfeasance. Some examples:

- In 1996, after the husband of a Maryland woman, Sharon R. Lopatka, reported her missing and said she had left a note saying she went to Georgia to visit friends and would not be coming back, police commenced a search that led them to her computer emails. There they found hundreds of electronic messages she exchanged with a man who called himself "Slowhand," whom they identified as Robert Glass. Some of the messages contained torture and death

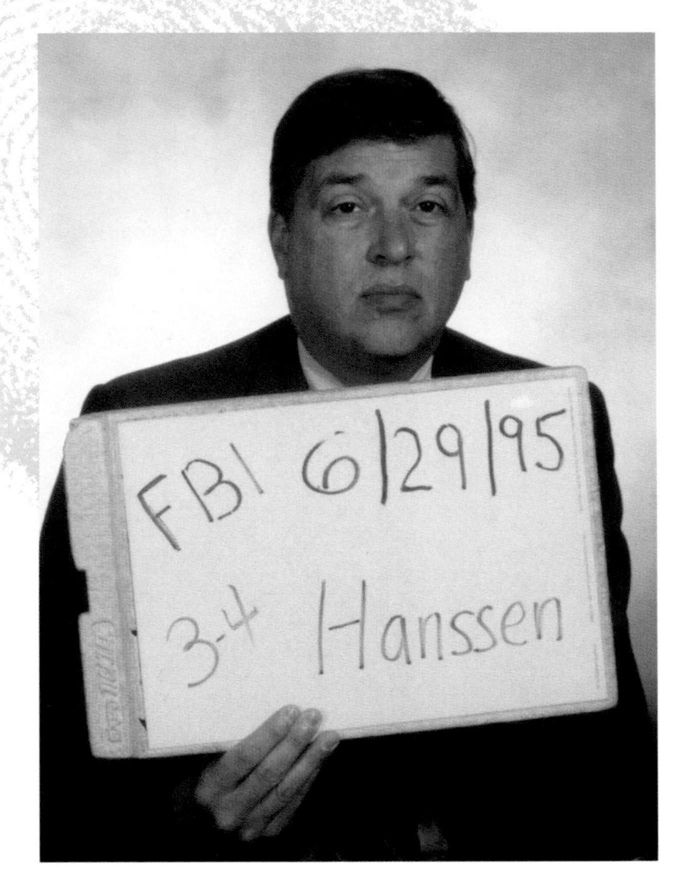

On the 20th anniversary of his service in the FBI, this amusing picture was taken for convicted spy Robert Hanssen's personal family album. Hanssen had by then been in the pay of the Russians for 10 years. One of his very early betrayals in 1985 led to the execution of two "turned" Soviets.

fantasies. Following the cyber trail to Glass in Collettsville, North Carolina, police ended up exhuming a suspicious area behind his trailer park to find Lopatka's battered body. An autopsy determined she had been sexually tortured and strangled three days earlier. Digital evidence had led the way.

- In 2001, federal agents arrested Robert P. Hanssen, a longtime FBI counter-intelligence agent, who had turned into a spy-for-hire against the United States. Part of the evidence against him was gleaned from his Palm III electronic device, which he had used to keep track of his schedule to pass information to his Russian contacts. (He was in the process of switching to a Palm VII to take advantage of its wireless capabilities.)
- Despite massive document shredding that had eliminated much of the paper trail needed to prove monumental fraud and theft by Houston's politically powerful Enron Corporation, forensic specialists were able successfully to recover many incriminating documents from the electronic backup files that had been overlooked by the accountants. This digital evidence helped to provide the smoking gun.
- In a Texas homicide case, the perpetrator turned out to be a person who was listed on the victim's handheld electronic organizer.
- In California, police investigated the high-profile murder of a seven-year-old girl, Danielle van Dam, who was snatched from her home near San Diego. Part of the probe involving the cops' seizure of four computer hard drives and a Palm Pilot belonging to a lead suspect, David A. Westerfield. As investigators tried to determine whether Westerfield or his son had viewed particular pornographic images on a computer, however, they reached some mistaken

conclusions. For example, one forensic examiner did not realize that the time-and-date stamps on an important email were in GMT rather than local time. But prosecutors ended up gaining Westerfield's capital conviction.

- On Virginia's Appalachian Trail, somebody discovered the decomposing body of an unidentified male who had been shot. Police suspected suicide and the clue leading to his identification proved to be a hand-held electronic device that contained the name of a 55-year-old Maryland man. Investigation determined he was the deceased and the death was indeed ruled a suicide.
- In Jacksonville, Florida, two men riding all-terrain vehicles in the woods discovered a woman's partially buried, dismembered torso and called the police. Investigators combed the crime scene and found a small piece of a pricing label that they identified as having come from a type of shovel sold only at Home Depot stores. With help from the store chain's security staff, they used the bar code from the shovel to determine the location and time of sale, and then viewed the corresponding surveillance video. The video produced images of a suspect who was identified as the victim's former boyfriend, who confessed to the murder.
- In St. Louis, Missouri, police reported they had linked six murders of prostitutes with DNA from semen and entered the profile into CODIS (Combined DNA Information System), but failed to turn up any matches. After a story about the cases appeared in the *St. Louis Post Dispatch*

The Defense cross-examines San Diego detective James Tomsovic during the trial of David Westerfield. He talks through pictures of Westerfield's motorhome. One of the computers found there is prominently displayed in the photographic evidence in the background.

David Westerfield on trial, June 18, 2002, listens to the opening forensic evidence; computer forensics examiner James Watkins found pornographic images of females who appeared to be under 18 on the computers of the accused.

the newspaper received an anonymous letter that said human remains for "victim number 17" were located within "a 50-yard radius from the X" that was inscribed on an accompanying computer-generated map of the West Alton area. The newspaper gave the map to police, who used it to find skeletal remains, and determined the spot was within a few hundred yards of where other bodies had been hidden. The accompanying map proved to be the evidence that cracked the case. After St. Louis police enlisted the assistance of the Illinois State Police Cyber Crime Unit and the FBI. Agents tracked the origin of the map to the Web site of Expedia and approached the company with a federal subpoena calling for it to provide the IP addresses of all computers that had recently visited the Web site. The search found there was only one user during that period. The Internet service provider divulged who was assigned to the IP address and police went after Maury Roy Travis. Records indicated he was a convicted felon and police obtained a search warrant for his home. They quickly found blood smears and other physical evidence to indicate the site was linked to the torture-murders of several women. Videotapes found hidden in the walls documented each of his sadistic killings.

Without digital evidence many cases would never have been solved.

POTENTIAL DIGITAL EVIDENCE IN HOME OR OFFICE

Computer Systems

Computer files: Evidence is most commonly found in files that are stored on hard drives and storage devices and media. Some examples may include user-created files, user-protected files, computer-created files such as backup files, history files, hidden files, printer spool files, deleted files.

Central Processing Units: The CPUs or microprocessor device itself may be evidence of component theft, counterfeiting, or remarking.

Servers: Gateway into computer files.

Answering Machine May store voice messages with time and date information about when the message was left.

Digital Cameras May hold images, removable cartridges, sound, time and date stamp, video.

Handheld Devices (Personal Digital Assistants, Pagers, and Electronic Organizers) May contain address book, appointment calendars/information, documents, email, handwriting, phone book, text, or voice messages.

Printer May maintain usage logs, time and date information, identity information, documents, hard drive, ink cartridges, superimposed images on the roller, time and date stamp.

Scanner The device itself may be evidence. Having the capability to scan may help prove illegal activity (e.g., child pornography, check fraud, counterfeiting, identity theft). Scanners can also leave unique marks.

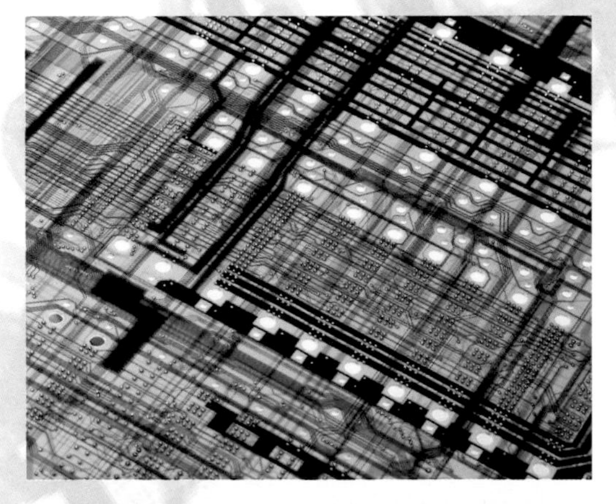

Telephone Can store names, phone numbers, and caller ID, may also store appointment information, receive electronic mail and pages, record voice messages.

Copier May contain documents, time and date stamp, user usage log.

Credit Card Skimmer Cardholder information contained on the tracks of the magnetic stripe includes card expiration date, credit card numbers, user's name and address.

Digital Watch May contain address book, email, notes.

Fax Machines May hold documents, film cartridge, phone numbers, send/receive log.

Global Positioning System Indicates home, previous destinations, travel logs, waypoint coordinates, waypoint name.

DIGITAL EVIDENCE SOUGHT FOR VARIOUS CRIMES

Auction Fraud (Online)

Account data regarding online auction sites.

Accounting/bookkeeping software and associated data files.

Address books.

Calendar.

Chat logs.

Customer information/credit card data.

Databases.

Digital camera software.

Email/notes/letters.

Financial/asset records.

Image files.

Internet activity logs.

Internet browser history/cache files.

Online financial institution access software.

Records/documents of "testimonials."

Telephone records.

UK killer of at least 215 of his elderly patients— probably many more—the arrogant Dr. Harold Shipman was too stupid to see that altering patient computer records retrospectively would leave a trail.

Child Exploitation/Abuse

Chat logs.

Date and time stamps.

Digital camera software.

Email/notes/letters.

Games.

Graphic editing and viewing software.

Images.

Internet activity logs.

Movie files.

User-created directory and file names that classify images.

Computer Intrusion

Address books.

Configuration files.

Email/notes/letters.

Executable programs.

Internet activity logs.

Internet protocol (IP) address and user name.

Internet relay chat (IRC) logs.

Source code.

Text files (user names and passwords).

Death Investigation

Address books.

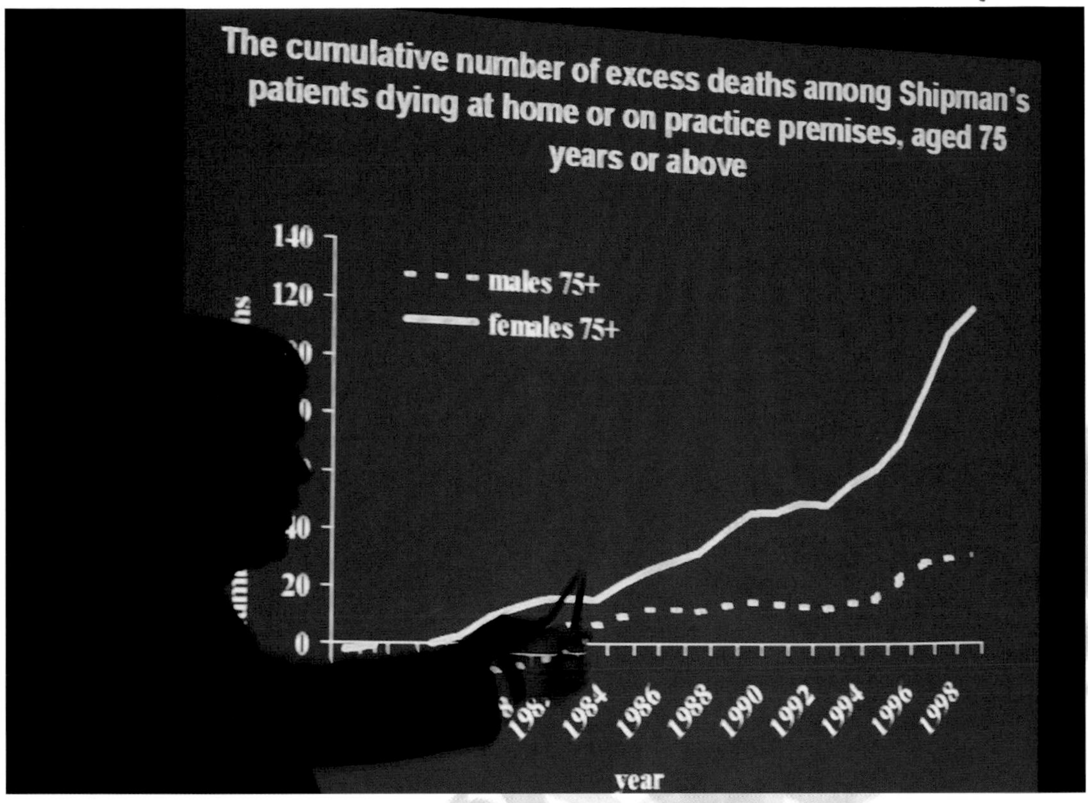

Professor Richard Baker, Professor of Quality in Health Care and Director of the Clinical Governance R & D Unit, Leicester University, delivers his report on Shipman, January 5, 2001. His forensic statistical work pointed to as many as 265 victims.

Diaries.
Email/notes/letters.
Financial/asset records.
Images.
Internet activity logs.
Legal documents and wills.
Medical records.
Telephone records.

Domestic Violence
Address books.
Diaries.
Email/notes/letters.
Financial/asset records.
Medical records.
Telephone records.

Economic Fraud (Including Online Fraud, Counterfeiting)
Address books.
Calendar.
Check, currency, and money order images.
Credit card skimmers.
Customer information/credit card data.
Databases.
Email/notes/letters.
False financial transaction forms.
False identification.

Financial/asset records.
Images of signatures.
Internet activity logs.
Online financial institution access software

Email Threats/Harassment/Stalking
Address books.
Diaries.
Email/notes/letters.
Financial/asset records.
Images.
Internet activity logs.
Legal documents.
Telephone records.
Victim background research.

Extortion
Date and time stamps.
Email/notes/letters.
History log.
Internet activity logs.
Temporary Internet files.
User names.

Gambling
Address books.
Calendar.

Customer database and player records.
Customer information/credit card data.
Electronic money.
Email/notes/letters.
Financial/asset records.
Image players.
Internet activity logs.
Online financial institution access software.
Sports betting statistics.

Identity Theft
Hardware and software tools.
- Backdrops.
- Credit card generators.
- Credit card reader/writer.
- Digital cameras.
- Scanners.

Identification templates.
- Birth certificates.
- Check cashing cards.
- Digital photo images for photo identification.
- Driver's license.
- Electronic signatures.
- Fictitious vehicle registrations.
- Proof of auto insurance documents.
- Scanned signatures.
- Social security cards.

Internet activity related to ID theft.
- Emails and newsgroup postings.
- Erased documents.
- Online orders.
- Online trading information.
- System files and file slack.
- World Wide Web activity at forgery sites.

Negotiable instruments.
- Business checks.
- Cashiers checks.

Counterfeit money.
Credit card numbers.
Fictitious court documents.
Fictitious gift certificates.
Fictitious loan documents.
Fictitious sales receipts.
Money orders.
Personal checks.
Stock transfer documents.
Travelers checks.
Vehicle transfer documentation.

Narcotics
Address books.
Calendar.
Databases.
Drug recipes.
Email/notes/letters.
False identification.
Financial/asset records.
Internet activity logs.
Prescription form images.

Prostitution
Address books.
Biographies.
Calendar.
Customer database/records.
Email/notes/letters.
False identification.
Financial/asset records.
Internet activity logs.
Medical records.
World Wide Web page advertising.

Software Piracy
Chat logs.
Email/notes/letters.
Image files of software certificates.
Internet activity logs.
Serial numbers.
Software cracking information and utilities.
User-created directory and file names that classify copyrighted software.

Telecommunications Fraud
Cloning software.
Customer database/records.
Electronic Serial Number (ESN)/Mobile Identification Number (MIN) pair records.
Email/notes/letters.
Financial/asset records.
"How to phreak" manuals.
Internet activity.
Telephone records.

Computer Crime Scene

Handling digital evidence: experts scramble to develop guidelines and standards for the recovery, preservation, and examination of digital evidence, including audio, imaging, and electronic devices.
By the 1990s police agencies throughout the United States—and of course across the World—were faced with a steep learning curve regarding the use of computers and digital evidence. Efforts began to develop cross-disciplinary guidelines and standards for digital evidence. Much of this activity involved trying to apply basic forensic science principles that were tried and trusted

with regard to conventional physical evidence to computer system search and seizure. Some of the basic principles at work included the recognition that:

- Actions taken to secure and collect electronic evidence should not change that evidence.
- Persons conducting examination of electronic evidence should be trained for the purpose.
- Activity relating to the seizure, examination, storage, or transfer of electronic evidence should be fully documented, preserved, and available for review.

Many of these issues appeared rather formidable as the scientific and legal status of the field continued to evolve rapidly. For example, the requirement that chain-of-custody documentation must be maintained for all digital evidence, as for other types of forensic evidence, still needed to be understood and defined, given the ability of computers to copy and alter data and the ability to easily transfer various versions of text or images relating to a crime scene.

Research established that while electronic evidence is ubiquitous, its nature also poses special challenges for its admissibility in court. To meet these challenges, proper forensic procedures had to be developed for the collection, examination, analysis, and reporting of information and data of investigative value that is stored on or transmitted by an electronic device.

Some of the particular characteristics of digital evidence were identified as follows:

- Electronic evidence is often latent (invisible) in the same sense as fingerprints or DNA evidence.
- It can transcend borders with ease and speed.
- It is fragile and can be easily altered, damaged, or destroyed.
- It is sometimes time-sensitive.

In view of these prerequisites, experts determined that precautions must be taken in the collection, preservation, and examination of electronic evidence. Computer forensic specialists had to be taught how to recognize and identify potential evidence; how to document the crime scene; how to handle such fragile material in ways that would not alter the evidence; how to know what special tools, equipment and techniques must be used to collect electronic evidence, particularly in view of constantly changing technologies; how a first responder should evaluate the scene and formulate a search plan; and so on.

One of the challenges associated with electronic evidence is that some data found on pagers, caller ID boxes, electronic organizers, cell phones, and other similar devices, is perishable. The first responder should always keep in mind that any device containing perishable data should be immediately secured, documented, and/or photographed.

Keyboards, the computer mouse, diskettes, CDs, or other components may have latent fingerprints or other physical evidence linking it to a suspect or victim and this evidence should be preserved. Chemicals used in processing latent prints can damage equipment and data.

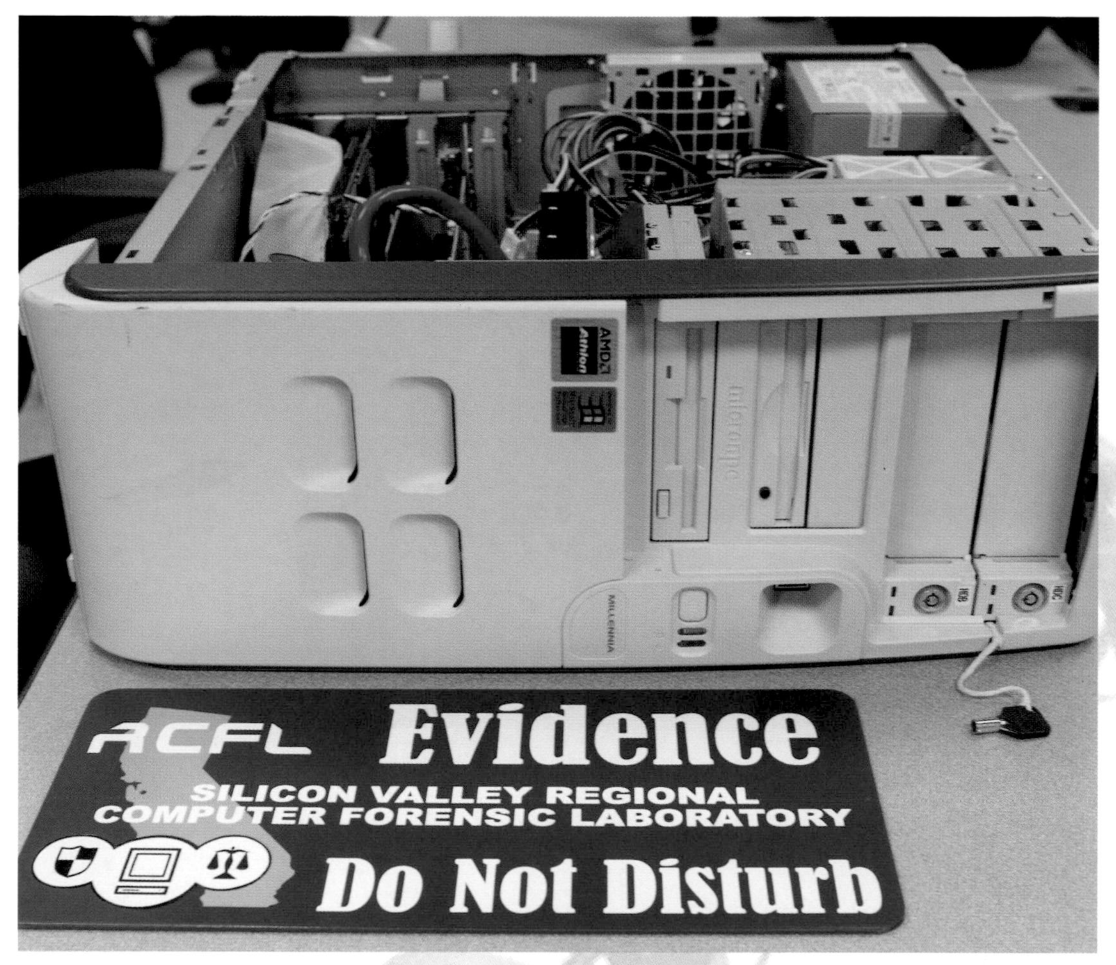

A computer opened on an examination table at the FBI's Silicon Valley Regional Computer Forensics Laboratory, CA, 2005. The new multimillion dollar facility was a recognition that digital evidence was swiftly growing in importance and that protocols were required. The laboratory is supposed to act as a kind of guarantor for the courts that the evidence is bone fide and uncontaminated.

Therefore, latent prints should be collected after the electronic evidence recovery is complete.

Investigation should be conducted in an effort to determine who are the owners and/or users of electronic devices found at the scene, as well as passwords, user names, and Internet service provider. As with other types of forensic evidence, documentation of the crime scene involving electronic evidence creates a permanent historical record that may prove important in the case. Investigators are trained to observe and document such details as the position of the mouse and the location of computer system components relative to each other, because, for example, a mouse on the left side of the computer may indicate a left-handed user. They are also expected to document the condition and location of the computer system, including power status of the computer (on, off, or in sleep mode). Most computers have status lights that indicate the computer is on. Likewise, if fan noise is heard, the system is probably on. If the computer system is warm, that may also indicate that it is on or was recently turned off. Photos and notes should describe what appears on the monitor screen.

Computer evidence, like all other evidence, must be handled carefully and in a manner that preserves its evidentiary value. This relates not just to the physical integrity of an item or device, but also to the electronic data it contains. Certain types of computer evidence, therefore, require special collection, packaging, and transportation. Consideration should be given to protect data that may be susceptible to damage or alteration from electromagnetic fields such as those generated by static electricity, magnets, radio transmitters, and other devices.

Recovery of non-electronic evidence can be crucial in the investigation of electronic crime. Proper care should be taken to ensure that such evidence is recovered and preserved. Items relevant to subsequent examination of electronic evidence may exist in other forms (e.g., written passwords and other handwritten notes, blank pads of paper with indented writing, hardware and software manuals, calendars, literature, text, or graphical computer printouts, and photographs) and should be secured and preserved for future analysis. These items frequently are in close proximity to the computer or related hardware items. All evidence should be identified, secured, and preserved in compliance with departmental policies.

Prior to starting the evidence collection phase, computer forensic investigators need to determine if the devices are individual, stand-alone desktops or laptops, or part of a computer network, because each of these requires different procedures. Securing and processing a crime scene where the computer systems are networked poses special problems because improper shutdown may destroy data. This can result in loss of evidence and potential severe civil liability. When investigating criminal activity in a known business environment, the presence of a computer network should be planned for in advance, if possible, and appropriate expert assistance obtained. Computer networks can also be found in a home environment and the same concerns exist.

The important forensic evidence may be inside the computer—but the hardware itself is still important. A photographic record is a priority, as for any crime scene: using a mouse to catch a rat.

Computers are fragile electronic instruments that are sensitive to temperature, humidity, physical shock, static electricity, and magnetic sources. Therefore, special precautions should be taken when packaging, transporting, and storing electronic evidence. To maintain chain of custody of electronic evidence, document its packaging, transportation, and storage. Any improper touching or moving of a computer can ruin its evidentiary value, so crime scene investigators are trained to exercise extreme caution.

Other types of electronic devices, such as facsimile machines, handheld recorders, and such, all carry their own particular requirements, challenges and advantages. For instance, when a cell phone is found, proper forensic protocol may call for it to be placed in a Faraday bag, which prevents stray signals from going in or out. The bag is made of metallic mylar, with a small metal grid on the surface. If such a bag is not available, putting the phone in an empty paint can and closing the lid will also work.

In addition to the usual forensic tools such as cameras, notepads, evidence forms, etc., computer forensic tools can include some not so obvious equipment. The lowly cable tag and stick-on label is a must. Screwdrivers are need for disassembly, including specialized screwdrivers from manufacturers such as Compaq or Macintosh, plus wire cutters, tweezers, etc.

Alongside the usual evidence bags, antistatic bags are essential, plus packing materials that won't produce static—so Styrofoam is no good. Unused diskettes are a must, as is printer paper, plus maybe a hand truck for taking away hardware.

HIDDEN EVIDENCE

In some cases, such as large-scale drug trafficking, white-collar crime, espionage, or terrorism, criminals may go to extraordinary lengths to conceal their computerized data. Encryption and other security methods can make it difficult for anyone but an approved user to access.

In some instances, the criminal may even adopt procedures to hide the messages by means of steganography, which is the science of writing hidden messages in such a way that no one apart from the recipient knows of the existence of the message. This is unlike cryptography, because an encrypted message does not actually hide, it's just in code. Large digital pictures, for example, may contain hidden text embedded in some of the bits. Accessing this protected or hidden information can prove too daunting for all but the most proficient computer forensics specialists. How often this encoding is done, however, is open to question. There were rumors in the press post 9/11 that terrorists were imparting information to each other steganographically in digital images on eBay. This was dismissed by experts, but that didn't stop the public scanning millions of images for steganographic activity. None was ever found.

The evaluation of the digital data seized in complex cases is the greatest challenge of all: what can be made of the data? Investigations into the 9/11 terrorist attacks, for example, yielded an estimated 125 terabytes of data, or many times more than the contents of the Library of Congress. Finding a "smoking gun" amid such vast reaches of data can require a lot more than speed reading—it requires the world's best search engines.

Digital Imaging

Does today's pervasive digital imaging technology come with ingrained problems for forensic science?

Two embarrassing incidents involving the FBI's massive Integrated Automated Fingerprint Identification System (IAFIS) left some observers wondering whether inherent flaws of digital imaging technology, not human error, might have been responsible. Established in 1999, IAFIS has grown to become the world's largest collection of human fingerprints, which are now checked almost instantaneously.

In 2004 the FBI apologized to Oregon lawyer Brandon Mayfield for accusing him of involvement in the Madrid terrorist bombing based on faulty fingerprint matching. The Bureau promised to review its current practices and convene an international panel of fingerprint experts to determine what went wrong.

A year later, the agency admitted that it had missed a fingerprint match for a man, Jeremy B. Jones, who had been arrested in Georgia on a minor charge. The database should have showed he had an outstanding warrant on serious charges, but the lack of any such match resulted in his release. He then went on to kill three women and one teenage girl in three states.

According to the FBI, in the Mayfield case Spanish authorities submitted to the FBI digital images of partial latent fingerprints found on blue plastic bags that contained detonator caps used in the train bombing. These images were sent through IAFIS, which is built on digital imaging technology. An IAFIS search compares an unknown fingerprint to a database containing known prints. The result of an IAFIS search can produce a short list of potential matches. A trained fingerprint examiner is then supposed to carefully compare the short list of these potential matches to the unknown print to see if it positively identifies a specific individual.

Muslim attorney Brandon Mayfield is released after two weeks detention. Erroneous fingerprint analysis in Spain had implicated him in the Madrid attacks. The FBI apologised.

Using this new methodology, a partial latent fingerprint from the Madrid attack was linked to Mr. Mayfield's fingerprint in the database, according to federal authorities.

In the wake of the Jones case, the FBI acknowledged that IAFIS was "an exceptional tool for law enforcement," but "not perfect." Although an agency spokesman attributed the error to "a result of a technical database error," some scientists worried that the problem may run much deeper—to the very nature of digital imaging. The first case resulted in a false positive, and concerns arose that the second blooper might have involved a false negative.

The sudden proliferation of digital technology has revolutionized law enforcement, equipping police officers with digital cameras to photograph crime scenes, digital video units to conduct surveillance and capture crimes in progress, scanners and printers to transfer fingerprints and other images, computer systems to store and retrieve data, digital fax machines, and software to create and edit digital images, giving even small police departments their own digital forensic laboratories to solve crimes and apprehend criminals. Almost overnight, digital photography has practically replaced analog film. Part of the reason why digital gadgetry has proven so popular is its relatively low cost and the tremendous speed and ease with which it can be used to carry out multiple tasks.

But some fundamental problems with the technology may have been overlooked. Digital technology relies on Charged Couple Devices (CCD's) that represent color by averaging the colors near a given light receptor. By their very nature, CCD's degrade detail and make the color less accurate than ordinary (analog) film. When a multiplicity of such devices—cameras, fax machines, scanners—are interconnected and used in a chain, the original image becomes even more distorted.

On top of this problem, commonly available software also enables users to manipulate digital images to "enhance" their quality and make them more identifiable or distinct. Detecting how the original has been altered may later prove impossible. In other words, digital technology includes some inherent pitfalls for forensic science.

Thus far, the courts have generally accepted digital evidence, but that can change, according to Edward Imwinkelried, a law professor at the University of California at Davis and a leading national expert on criminal evidence. A federal judge in Portland threw out the criminal case against Mr. Mayfield after federal agents acknowledged his fingerprint didn't match the mark left in Madrid. But the technology itself has remained generally acceptable to the courts.

Imwinkelried expressed particular concern about the application of digital technology in forensic science. "The fact that the technology was validated for the space program doesn't mean it can be used in forensics," he said. "Another body of research is needed."

Michael Cherry, an imaging specialist from New Jersey who in the late 1980s helped to pioneer the replacement of film photography with digital electronic photography, warned that users of digital imaging needed to be more careful in scrutinizing such evidence. Legal doctrine requires that, to be produced as criminal evidence, the item must be the original. Yet digital enhancement subtracts pixels and changes the image, thereby making it different. One of the as-yet unresolved problems of the digital age is that so much crucial data relies on imagery of undetermined provenance and accuracy, and is susceptible to abuse. Some of these issues are now being studied by the National Academy of Sciences.

Postscript

The history of forensic science reveals many incredible advances and stunning setbacks that may offer a clue to the future.

Contrary to the Hollywood version, reality demands a more mixed and cautionary verdict on the role of forensic science in criminal investigation. Forensic science hasn't always held all of the answers, or managed to overcome every barrier in its path. The good guy hasn't always won. Looking back at some of the criminal-investigation methods—and the state of the underlying knowledge in many scientific fields on which they were based—can produce as much skepticism as wonder about forensic evidence in the past. Considering that what was done before shapes the present and influences the future, forensic enthusiasts may find ample reason to be wary.

On the brighter side, science is always advancing. And forensic sciences are no exception. Indeed, the march of progress in the forensic field has lately been extremely impressive, owing particularly to the harnessing of computers, digital imaging, DNA profiling, and other technological improvements. The expansion of crime laboratories, enhanced field-testing capabilities, and increased training of criminalists and other forensic personnel has reduced processing times and boosted productivity. Courts have become more discriminating in their consideration of scientific evidence. Specialties within disciplines have appeared, fostering more research studies, validation studies, knowledge exchange, peer review, and internal policing. As a result, "junk science" has come under greater scrutiny. The movement toward greater professionalism, certification, and accreditation has also raised the performance level, providing better criteria by which to evaluate individual proficiency and institutional performance.

The wonders of DNA singularity have often enabled scientists to make astonishing identifications of murderers, rapists, and other serious offenders who otherwise may have eluded capture. DNA has also proved nothing short of miraculous in helping police to identify victims of mass disasters and individual, unexplained death. At the same time, DNA has exposed wrongful convictions and bad arrests. In so doing, it has revealed serious flaws in confessions, eyewitness identification, bite mark identification, fingerprint evidence, and many other common practices that previously seemed pretty solid or even infallible.

In the process, the DNA revolution has stimulated unprecedented concern about chain of custody, evidence contamination, and evidence fabrication—concern that some savvy criminals have also sought to exploit, in some instances by planting another individual's DNA at a crime scene. It has caused not only many other kinds of evidence but also the experts who handle them to undergo newfound scrutiny. For good reason.

Because forensic science by its very nature must operate in the context of law, and in view of many inevitable conflicts between legal and scientific approaches, the effectiveness of the field must ultimately be judged—rightly or wrongly—in courts of law, not in scientific journals. The history of the evolution of forensic science in the United States reflects changing political notions about the role of law in American society as much as it mirrors breakthroughs in science.

If the past offers any guide, we can scarcely imagine what the future may hold. Terrorist crimes such as the Oklahoma City Bombing, 9/11, and the Anthrax attacks, and the crime scene investigation challenges they pose may pale in comparison to the horror of tomorrow's crimes involving hazardous chemicals, biological materials, and radioactive isotopes. Evidence gathering relative to use of such WMD and toxic warfare has already begun to create a new field, known as "hot zone forensics." Weapons of mass destruction and toxic warfare weapons represent a threat of mind-boggling proportions, requiring methods that can only provide salvation if they can help to head off or prevent their ever being used in the first place. Crime scenes continue to hold many hidden clues about human relations, bearing witness to acts of the worst sort. The job of crime scene investigators and other forensic scientists is to dig them out and hold them up for the world to see, so there is no doubt about precisely how they were carried out, and who committed the crime. The question of what should be done in response is best left to others.

The author on the right examines the skull of a New York City murder victim. Dr. Michael M. Baden is second from left. In his long career, Dr. Baden has conducted more than 20,000 autopsies. Besides the high-profile cases already mentioned in this book, he has brought his expertise to bear on the death of John Belushi, Billy Martin, and the victims of TWA Flight 800.

Forensics Time Line

1932 FBI forensic science laboratory established

1935 Bruno Hauptmann convicted of kidnapping death of Lindbergh baby in New Jersey

1944 Hartford, Connecticut, circus tent fire; 162 of 268 bodies identified by dental records

1947 U.S. Army forms Central Identification Laboratory in Hawaii to assist in identifying military human remains

1950 American Academy of Forensic Sciences is founded in Chicago

1953 Paul Leland Kirk of University of California publishes Crime Investigation textbook in criminalistics

1954 Texas courts allows bite-mark evidence

1954 R.F. Borkenstein invents Breathalyzer for field sobriety testing

1957 Skeletal growth stages, the basis of forensic anthropology, identified by American pathologists Thomas Mocker and Thomas Stewart

1960 First laser design to identify fingerprints (and other applications) developed by U.S. physicist Theodore Maiman

1964 Warren Commission Report on John F. Kennedy assassination

1966 Dr. Sam Sheppard acquitted of murder after second trial in Ohio

1967 FBI National Crime Information Center (NCIC) established

1970 Forensic odontology division created in American Academy of Forensic Sciences

1971 Photo-fit developed by photographer Jacques Perry

1974 Detection of gunshot residue using scanning electron microscopy with electron dispersive X-rays

1977 FBI starts Automated Fingerprint Identification System (AFIS)

1978 Electro-Static Detection Apparatus (ESDA) to expose handwriting impressions developed by Bob Freeman and Doug Foster

1979 Bite evidence key to convicting serial killer Ted Bundy in Florida; dental records help to identify 913 victims of Jim Jones cult in Jonestown, Guyana

1980 Method for detecting DNA differences developed by American Ray White

1982 Wayne Williams convicted in Atlanta child murders case based on fiber/hair evidence

1983 First use of personal computers in U.S. police patrol cars to provide quick information from National Crime Information Center

1984 Genetic profiling using DNA developed by English geneticist Professor Alec Jeffreys

1985 First paper published about polymerase chain reaction (PCR)

1987 First time DNA evidence used to get a conviction in a U.S. criminal court with Tommy Lee Andrews sexual assaults case in Orlando, Florida

1989 Conviction of Richard Ramirez in California Night Stalker case

1989 Illinois convicted rapist Gary Dotson becomes first person exonerated by DNA

1993 *Daubert et al. v. Merrill Dow* replaces *Frye* standard for admission of scientific evidence

1993 Seventy-five die in Branch Davidian compound, Waco, Texas

1993 Terrorists bomb the World Trade Center, NYC

1994 Convicted in 1990 of the rape of a patient in the Metropolitan State Hospital, Los Angeles, Mark Diaz Bravo is released following DNA testing of items from crime scene

1995 Oklahoma City federal building bombing

1995 O.J. Simpson is acquitted of murder in Los Angeles

1996 Digital dental radiography used to identify human remains in air crash of TWA Flight 800 off Long Island

1996 Unabomber Theodore Kaczynski, who terrorized the nation from 1978–95, killing three persons and injuring at least 23 others, is arrested after a tip from his brother

1998 FBI launches its DNA database, NIDIS

1999 FBI implements Integrated Automated Fingerprint Identification System (IAFIS)

2001 Attack on the World Trade Center and Pentagon triggers war on terrorism with forensic science playing a pivotal role

2001 Anthrax terrorist strikes in U.S. mails

2002 DC snipers terrorize Washington-Maryland beltway

2005 Former FBI agent Jeffrey Royer and Internet penny stock advisor Anthony Elgindy convicted for mining confidential government computer files to manipulate stocks

New York City Police Department photograph of (from left to right) "Waxie" Gordon, Hymie Pincus, and Albert Aront, May 21, 1933. Waxie Gordon was right up there with "Dutch" Schultz, "Lucky" Luciano, Frank Costello, and Al Capone as a bootlegging millionaire. Since its invention, the camera has been a vital tool in criminal investigation and still is. But as pointed out earlier (see page 172), today's digital imagery poses particular problems for forensics because of its "malleability."

Leading Forensics Associations and Societies

American Academy of Forensic Sciences (AAFS)
American Board of Criminalistics (ABC)
American Board of Forensic Anthropology (ABFA)
American Board of Forensic Odontology (ABFO)
American Board of Forensic Toxicology (ABFT)
American Society of Crime Laboratory Directors (ASCLD)
American Society for Quality (ASQ)
American Society of Questioned Document Examiners (ASQDE)
American Society for Testing and Materials (ASTM)
Association of Firearm and Tool Mark Examiners (AFTE)
Association of Forensic Quality Assurance Managers (AFQAM)
Forensic Science Society (FSS)
International Association for Identification (IAI)
International Association of Arson Investigators (IAAI)
International Association of Bloodstain Pattern Analysts (IABPA)
National Association of Medical Examiners (NAME)
National Center of Forensic Science (NCFS)
National Forensic Science Technology Center (NFSTC)
Society of Forensic Toxicologists (SOFT)
International Association of Forensic Toxicologists (TIAFT)

In 1977 a three-year federal study of 240 laboratories operated by federal, state, and local law enforcement agencies found many made glaring mistakes. The "blind" study guaranteed anonymity for all of the labs and avoided offering any conclusions about the impact of their shoddy work. The study found most labs were understaffed, under-trained, under-equipped, and under-financed. Ten years later, the DNA technology revolution prompted intense national concern about the professionalism and impartiality of laboratory forensic scientists. Part of the controversy arose from a scandal involving Joyce Gilchrist, a forensic chemist for the Oklahoma City Police Department, who fabricated lab results. The Federal Aviation Authority lab former supervisor of forensic toxicology was found to have falsified drug test results. Critics complained that the forensic field lacked any accreditation requirements, advanced academic degree programs, and recognized standards for professional performance. Since then, funding has increased dramatically and the associations above have underwritten a new professionalism.

Poisons and Their Symptons

Caustic Poison (lye)	Burns around victim's lips and mouth
Carbon Monoxide	Red or pink patches on chest and thighs
Sulfuric Acid	Black vomit
Hydrochloric Acid	Greenish-brown vomit
Nitric Acid	Yellow vomit
Silver Salts	White vomit turning black in daylight
Copper Sulfate	Blue-green vomit
Phosphorous	Coffee-brown vomit, onion or garlic odor
Cyanide	Burnt almond odor in air
Ammonia, Vinegar	Characteristic odors
Arsenic, Mercury, Lead Salts	Pronounced diarrhea
Methyl (Wood) Alcohol	Nausea and vomiting, blindness, unconsciousness
Isopropyl (Rubbing) Alcohol	Dizziness, CNS depression, nausea and vomiting, unconsciousness

SOURCE: U.S. Department of Justice, *Crime Scene Search & Evidence Handbook* (1973).

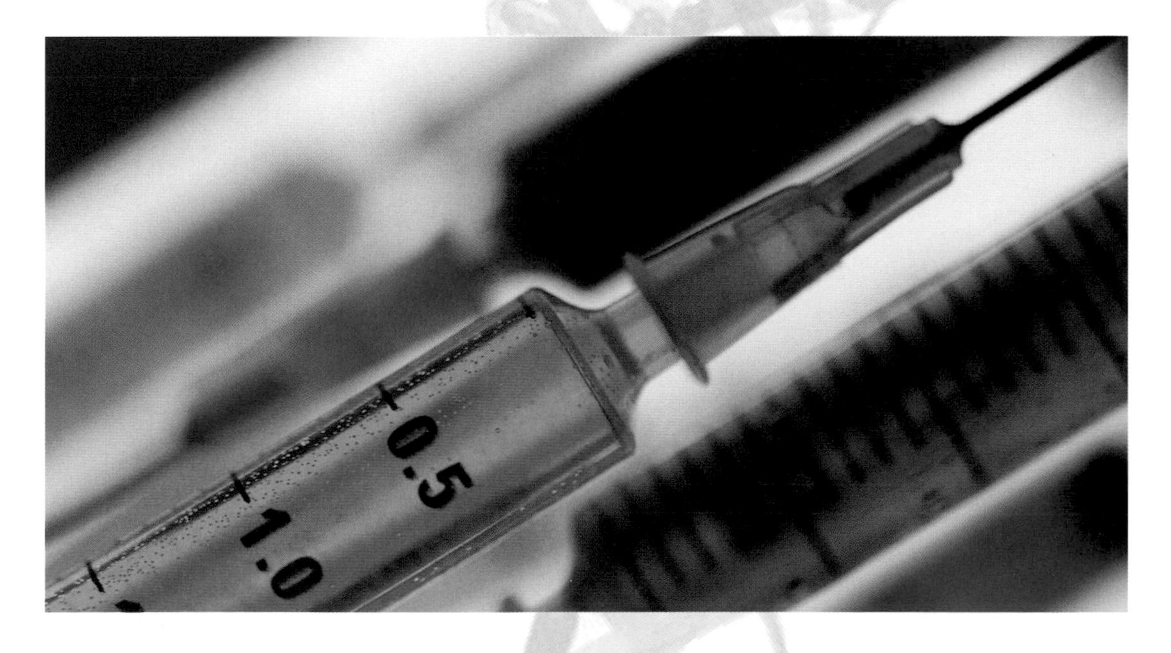

Crime Scene Investigation Glossary

ABFO scales. Measurement scales established by the American Board of Forensic Odontology in the form of an L-shaped piece of plastic used in photography that is marked with circles, black-and-white bars, and 188-percent gray bars to assist in distortion compensation and provide exposure determination. The plastic piece is marked in millimeters.

Alternate light source. Equipment used to produce visible or invisible light to enhance or visualize potential items of evidence such as fluids, fingerprints, or clothing fibers.

Biohazard bag. A container for materials that have been exposed to blood or other biological fluids and which therefore may be contaminated with AIDS, hepatitis, or other viruses.

Biological fluids. Fluids such as blood, mucus, perspiration, saliva, semen, vaginal fluid, urine or feces, that have a human or animal origin.

Biological weapon. Biological agents such as anthrax, smallpox or other infectious diseases that can be used as a life-threatening weapon of war or crime.

Bloodborne pathogen. Infectious, disease-causing microorganisms that may be found or transported in biological fluids.

Boundaries. The perimeter or borders that are established by the police to surround and protect the potential physical evidence related to a crime.

Bunter markings. The etching of a bullet's manufacturer and caliber on a casing.

Case file. The collection of official documents and other evidence comprising information concerning a particular criminal investigation.

Case identifiers. The alphabetic and/or numeric characters assigned to identify a particular criminal case.

Castoff. A bloodstain pattern created when blood is released from a blood-bearing object in motion.

Chain of custody. A careful process used to maintain and document the chronological history of the physical evidence. Such documentation includes the names or initials of the individual collecting the evidence, the identify of every person or entity who subsequently has custody of it, the dates the items were collected or transferred, the victim's or suspect's name or other case identifier, and a brief description of the item.

Chemical enhancement. The use of special, approved chemicals that react with specific types of evidence to aid in the detection or documentation of criminal evidence that may otherwise be difficult to see.

Chemical threat. Compounds that pose bodily harm if touched, breathed, ingested or ignited. Such compounds may be encountered at the scene of an explosion, tanker leakage, or other discharge.

Clean/sanitize. The process of removing biological or chemical contaminants from tools or equipment using approved methods, such as washing with a mixture of 10-percent household bleach and water.

Comparison samples. A generic term used to describe physical material or evidence that is discovered at a crime scene that may be compared with samples taken from identified/known or unknown/questioned persons, tools and physical locations. Examples of three different basic types include standard/reference samples, control/blank samples, and elimination samples.

Contamination. The unwanted or unauthorized transfer of material from another source to a piece of physical evidence.

Cross-contamination. The unwanted transfer of material between two or more sources of physical evidence.

Documentation. Written notes, sketches, photographs, audiotaped or videotaped impressions that provided a detailed record of the crime scene resulting from the crime scene investigation.

Evidence identifiers. Tape, labels, containers, string tags or other means used to identify the evidence, the person collecting the evidence, the date the evidence was gathered, basic criminal offense information, and a brief description of the pertinent evidence.

First responders. Initial responding law enforcement officer(s) or other public safety official(s) or service provider(s) arriving at the crime scene prior to the arrival of the official lead investigator.

Forensic science. The application of science to the facts related to criminal and civil litigation.

Impression evidence. Objects or materials that have retained the characteristics of other objects that have been physically pressed against them.

Latent print. A print impression not readily visible, made by contact of the hands or feet with a surface resulting in the transfer of materials from the skin to that surface.

Lividity. Discoloration of the skin through gravitation of the blood after death. First visible 30–60 minutes after the heart stops beating.

Measurement scale. An object (usually a rule) showing accurate units of length used in photographic documentation of an item of evidence.

Multiple scenes. Two or more physical locations of evidence associated with a crime. For example, a violent crime may include multiple locations such as the spot where the attack took place, the assailant's vehicle or home, or the victim's home.

Nonporous container. Packaging such as glass jars or metal containers through which liquids or vapors cannot pass.

Personal protective equipment. Articles such as disposable gloves, masks, or glasses used to protect crime scene investigators from biological or chemical hazards and to avoid contamination of the crime scene.

Porous container. Packaging such as paper bags or cloth sacks through which liquids or vapors may pass.

Presumptive test. A nonconfirmatory test used to screen for the presence of a substance.

Projectile trajectory analysis. The method of determining the path of a high-speed object through space, such as the course of bullets fired from a gun.

Single-use equipment. Items such as tweezers, droppers, or scalpel blades that can be used only once to collect evidence before they are discarded.

Trace evidence. Physical evidence that results from the transfer of small quantities of materials such as hair, textile fibers, paint chips, gunshot residue particles, or other tiny items.

Transient evidence. Evidence that by its very nature or the conditions at the scene will lose its evidentiary value if not preserved and protected, as happens to blood in rain or tire impressions left in snow.

Walk-through. An initial assessment of the crime scene conducted by carefully walking through the area to observe the situation, recognize potential evidence, and estimate the resources that will be required to investigate the physical evidence. It also may describe a final survey that is conducted to ensure the crime scene has been effectively and completely processed.

SOURCE Adapted from the U.S. Department of Justice, *Crime Scene Investigation: A Guide for Law Enforcement* (Washington, D.C.: Office of Justice Programs, U.S. Department of Justice, 1995).

A fingerprint case; on the left are different types of powder, the brushes are in the middle, and the tape to lift the prints is on the right. How important is fingerprint evidence? Consider the statistics for Greenville County SC, 2004, population around 400,000: total crime scene calls—7,363; processed for prints—4,941; latent prints processed, scenes—2,686; latent fingerprint examinations—463,962; fingerprint matches—1,610. Extrapolate across the country: it's important.

Bibliography

Blood and DNA

Tom Bevel and Ross M. Gardner, *Bloodstain Pattern Analysis: With an Introduction to Crime Scene Reconstruction,* Boca Raton: CRC Press, 1997.

John M. Butler, *Forensic DNA Typing: Biology & Technology Behind STR Markers,* San Diego: Academic Press, 2001.

Dr. Henry C. Lee and Frank Tirnady, *Blood Evidence: How DNA is Revolutionizing the Way We Solve Crimes,* Cambridge: Perseus, 2003.

Herbert MacDonnell and Loraine Bralousz, *Flight Characteristic and Stain Patterns of Human Blood,* Washington, D.C.: U.S. Government Printing Office, 1971.

Barry Scheck, Peter Neufeld, and Jim Dwyer, *Actual Innocence: Five Days to Execution and Other Dispatches of the Wrongfully Convicted,* New York: Doubleday, 2000.

Chemical, Biological & Radiological

Steven C. Drielak, *Hot Zone Forensics: Chemical, Biological, and Radiological Evidence Collection,* Springfield, Illinois: Charles C. Thomas, 2004.

Crime Laboratories

Jami St. Clair, *Crime Laboratory Management,* San Diego: Academic Press, 2002.

Criminalistics/Forensics

Colin Evans, *The Casebook of Forensic Detection,* New York: John Wiley & Sons, 1996.

Paul L. Kirk, *Crime Investigation,* 1st ed., New York: John Wiley & Sons, 1953.

Henry C. Lee, Timothy Palmbach and Marilyn Miller. *Henry Lee's Crime Scene Handbook,* San Diego: Academic Press, 2001.

C.E. O'Hara and J.W. Osterburg, *An Introduction to Criminalistics; The Application of the Physical Sciences to the Detection of Crime,* New York: Macmillan, 1949.

J.W. Osterburg, *The Crime Laboratory; Case Studies of Scientific Criminal Investigation,* Bloomington: Indiana University Press, 1968.

David Owen, *Hidden Evidence: Forty True Crimes and How Forensic Science Helped Solve Them,* Buffalo: Firefly Books, 2000.

Richard Platt, *Crime Scene: The Ultimate Guide to Forensic Science,* New York: DK Publishing, 2003.

Richard R. Saferstein, *Criminalistics: An Introduction to Forensic Science,* 6th edition, Englewood Cliffs, NJ: Prentice Hall, 1998.

Cyril H. Wecht et al, *Crime Scene: Crack the Case with Real-Life Experts,* New York: Reader's Digest, 2004.

Dentistry

Ira A. Gladfelter, *Dental Evidence: A Handbook for Police,* Springfield, Illinois: Charles C. Thomas, 1975.

Irvin M. Sopher, *Forensic Denistry*, Springfield, Illinois: Charles C. Thomas, 1976.

Paul G. Stimson and Curtis A. Mertz, Editors. *Forensic Dentistry*, Boca Raton: CRC Press, 1997.

Digital Evidence

Eoghan Casey, *Digital Evidence and Computer Crime,* San Diego: Academic Press, 2004.

———*Handbook of Computer Crime Investigation: Forensic Tools and Technology,* San Diego: Academic Press, 2001.

Searching and Seizing Computers and Obtaining Electronic Evidence in Criminal Investigations, Washington, D.C.: U.S. Department of Justice, Computer Crime and Intellectual Property Section, March 2001. www.cybercrime.gov/searchmanual.htm.

Firearms & Ballistics

Vincent Di Maio, *Gunshot Wounds: Practical Aspects of Firearms, Ballistics, and Forensic Techniques,* 2nd ed., Boca Raton: CRC Press, 1999.

A.J. Schwoeble and David L. Exline, *Current Methods in Forensic Gunshot Residue Analysis,* Boca Raton: CRC Press, 2000.

Fingerprints & Impressions

D.R. Ashbaugh, *Quantitative-Qualitative Friction Ridge Analysis: An Introduction to Basic and Advanced Ridgeology*, Boca Raton: CRC Press, 2000.

Colin Beavan, *Fingerprints: The Origins of Crime Detection and the Murder Case that Launched Forensic Science*, New York: Hyperion, 2001.

W.J. Bodziak, *Footwear Impression Evidence*, New York: Elsevier, 1990.

B.C. Bridges, *Practical Fingerprinting*, New York: Funk & Wagnalls, 1963.

Christophe Champod, Chris Lennard, Pierre Margot, and Milutin Stoilovic, *Fingerprints and Other Ridge Skin Impressions*, Boca Raton: CRC Press, 2004.

Simon A. Cole, *Suspect Identities: A History of Fingerprinting and Criminal Identification*, Cambridge: Harvard University Press, 2001.

Henry C. Lee and R.E. Gaensslen, *Advances in Fingerprint Technology*, New York: Elsevier, 1991.

E. Roland Menzel, *Fingerprint Detection with Lasers*, 2nd ed., New York: Marcel Dekker, 1999.

Louise M. Robbins, *Footprints: Collection, Analysis, and Interpretation*, Springfield, Illinois: Charles C. Thomas, 1985.

Forensic Anthropology

Bill Bass and Jon Jefferson, *Death's Acre: Inside the Legendary Forensic Lab Where the Dead Do Tell Tales*, New York: G.P. Putnam's Sons, 2003.

Karen Ramey Burnes, *Forensic Anthropology Training Manual*, Upper Saddle River, New Jersey: Prentice Hall, 1999.

Emily Craig, Ph.D., *Teasing Secrets from the Dead: My Investigations at America's Most Famous Crime Scenes*, New York: Crown, 2004.

Mahmoud Y. El-Najjar and K. Richard McWilliams, *Forensic Anthropology: The Structure, Morphology, and Variation of Human Bone and Dentition*, Springfield, Illinois: Charles C. Thomas, 1978.

Wilton Marion Krogman, *The Human Skeleton in Forensic Medicine*, 3rd Printing, Springfield, Illinois: Charles C. Thomas, 1962.

Robert B. Pickering and David C. Bachman, *The Use of Forensic Anthropology*, Boca Raton: CRC Press, 1997.

Forensic Entolomogy

Bernard Greenberg and John Charles Kunich, *Entomology and the Law: Flies as Forensic Indicators*, Cambridge, England: Cambridge University Press, 2002.

M. Lee Goff, *A Fly for the Prosecution: How Insect Evidence Helps Solve Crimes*, Cambridge: Harvard University Press, 2000.

Jessica Snyder Sachs, *Corpse: Nature, Forensics, and the Struggle to Pinpoint Time of Death*, New York: Perseus Books Group, 2002.

Jason H. Byrd and James L. Castner, eds., *Forensic Entomology: The Utility of Arthropods in Legal Investigations*, Boca Raton: CRC Press, 2000.

Tz'u Sung, *The Washing Away of Wrongs: Forensic Medicine in Thirteenth-Century China*, edited by Brian McKnight, Michigan: University of Michigan Center for Chinese Studies, 1981.

Explosions

Alexander Beveridge, ed., *Forensic Investigation of Explosions*, Bristol, PA: Taylor & Francis, Inc., 1998.

Handwriting & Documents

James V.P. Conway, *Evidential Documents*, 3rd Printing, Springfield, Illinois: Charles C. Thomas, 1959.

Ron N. Morris, *Forensic Handwriting Identification: Fundamental Concepts and Principles*, San Diego: Academic Press, 2000.

Albert S. Osborn, *Questioned Documents*, 2nd ed., New York: Boyd Printing Company, 1946.

Medical Examiners

Michael Baden with Judith Adler Hennessee, *Unnatural Death: Confessions of a Medical Examiner*, New York: Random House, 1989.

Michael Baden and Marion Roach, *Dead Reckoning: The New Science of Catching Killers*, New York: Simon & Schuster, 2001.

Milton Helpern with Bernard Knight, *Autopsy: The Memoirs of Milton Helpern, the World's Greatest Medical Detective,* New York: St. Martin's, 1977.

Pathology & Medicine

G.G. Brogdon, *Forensic Radiology,* Boca Raton: CRC Press, 1998.

Werner U. Spitz and Russell S. Fisher, *Mediolegal Investigation of Death,* 2nd Printing, Springfield, Illinois: Charles C. Thomas, 1973.

Photography

Larry L. Miller, *Police Photography,* 3rd edition, Cincinnati, Ohio: Anderson Publishing, 1993.

David R. Redsicker, *The Practical Methodology of Forensic Photography*, New York: Elsevier, 1991.

Police Investigation

Miles Corwin, *Homicide Special: A Year with the LAPD's Elite Detective Unit,* New York: Henry Holt, 2003.

V.J. Geberth, ed., *Practical Homicide Investigation: Tactics, Procedures and Forensic Techniques,* 3rd edition, Boca Raton: CRC Press, 1996.

V.J. Geberth, *Sex-Related Homicide and Death Investigation: Practical and Clinical Perspectives,* Boca Raton: CRC Press, 2003.

Steve Jackson, *No Stone Unturned: The True Story of the World's Premier Forensic Investigators,* New York: Pinnacle, 2002.

James J., Skehan, *Modern Police Work Including Detective Duty,* Brooklyn, New York: R.V. Basuino, 1939.

Harry Soderman and John J. O'Connell, *Modern Criminal Investigation,* New York: Funk & Wagnalls, 1935.

Profiling

James Brussel, *Casebook of a Crime Psychiatrist,* New York: Grove Press, 1968.

Don DeNevi and John H. Campbell, *Into the Minds of Madmen: How the FBI Behavioral Science Unit Revolutionized Crime Investigation,* Amherst, NY: Prometheus Books, 2004.

John Douglas, and Mark Olshaker, *Mindhunter: Inside the FBI's Elite Serial Crime Unit,* New York: Scribner, 1995.

Ronald M. Holmes and Stephen T. Holmes, *Profiling Violent Crimes,* 3rd edition, Thousand Oaks, CA: Sage, 2002.

Gregg McCrary with Katherine Ramsland, *The Unknown Darkness: Profiling the Predators Among Us,* New York: Morrow, 2003.

Brent Turvey, *Criminal Profiling,* San Diego: Academic Press, 1999.

The Dr. Sheppard Case

Jack P. DeSario, William D. Mason, *Dr. Sam Sheppard on Trial: The Prosecutors and the Marilyn Sheppard Murder,* Kent, OH: Kent State University Press, 2003.

James Neff, *The Wrong Man: The Final Verdict on the Dr. Sam Sheppard Murder Case,* New York: Random House, 2002.

Bernard F. Conners, *Tailspin: The Strange Case of Major Call,* New York: British American Publishing, 2002.

Cynthia L. Cooper and Sam Reese Sheppard, *Mockery of Justice: The True Story of the Sheppard Murder Case,* Boston: Northeastern University Press, 1995.

Dr. Sam Sheppard, *Endure and Conquer: My 12-Year Fight for Vindication,* Cleveland: The World Pub Co., 1996.

Paul Leland Kirk, *Crime Investigation,* New York: John Wiley & Sons, 1974.

F. Lee Bailey with Harvey Aronson, *The Defense Never Rests,* New York: Stein and Day, 1971.

Interviewed for this book, Sam Reese Sheppard said he hopes that some day it will be possible to find more evidence that will prove once and for all that his father did not murder his mother.

Trace Evidence

Max M. Houck, ed., *Trade Evidence Analysis: More Cases in Mute Witnesses,* New York: Elsevier, 2004.

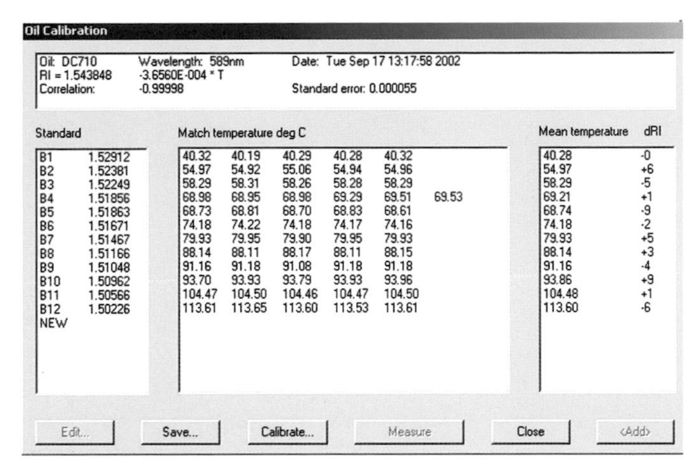

Oil Calibration

Oil: DC710 | Wavelength: 589nm | Date: Tue Sep 17 13:17:58 2002
RI = 1.543848 | -3.6560E-004 * T
Correlation: | -0.99998 | Standard error: 0.000055

Standard		Match temperature deg C						Mean temperature	dRI
B1	1.52912	40.32	40.19	40.29	40.28	40.32		40.28	-0
B2	1.52381	54.97	54.92	55.06	54.94	54.96		54.97	+6
B3	1.52249	58.29	58.31	58.26	58.28	58.29		58.29	-5
B4	1.51856	68.98	68.95	68.98	69.29	69.51	69.53	69.21	+1
B5	1.51863	68.73	68.81	68.70	68.83	68.61		68.74	-9
B6	1.51671	74.18	74.22	74.18	74.17	74.16		74.18	-2
B7	1.51467	79.93	79.95	79.90	79.95	79.93		79.93	+5
B8	1.51166	88.14	88.11	88.17	88.11	88.15		88.14	+3
B9	1.51048	91.16	91.18	91.08	91.18	91.18		91.16	-4
B10	1.50962	93.70	93.93	93.79	93.93	93.96		93.86	+9
B11	1.50566	104.47	104.50	104.46	104.47	104.50		104.48	+1
B12	1.50226	113.61	113.65	113.60	113.53	113.61		113.60	-6
NEW									

Edit... | Save... | Calibrate... | Measure | Close | <Add>

The memorably named GRIM (Glass Refractive Index Measurement) workstation. The refractive index of a glass fragment can be calibrated by immersing it in oil and then heating the oil until the RI of the oil and the glass are equal—i.e. there is no refraction, or bending of light passing through oil to glass. If the temperature at that point is the same for two samples, there is a match.

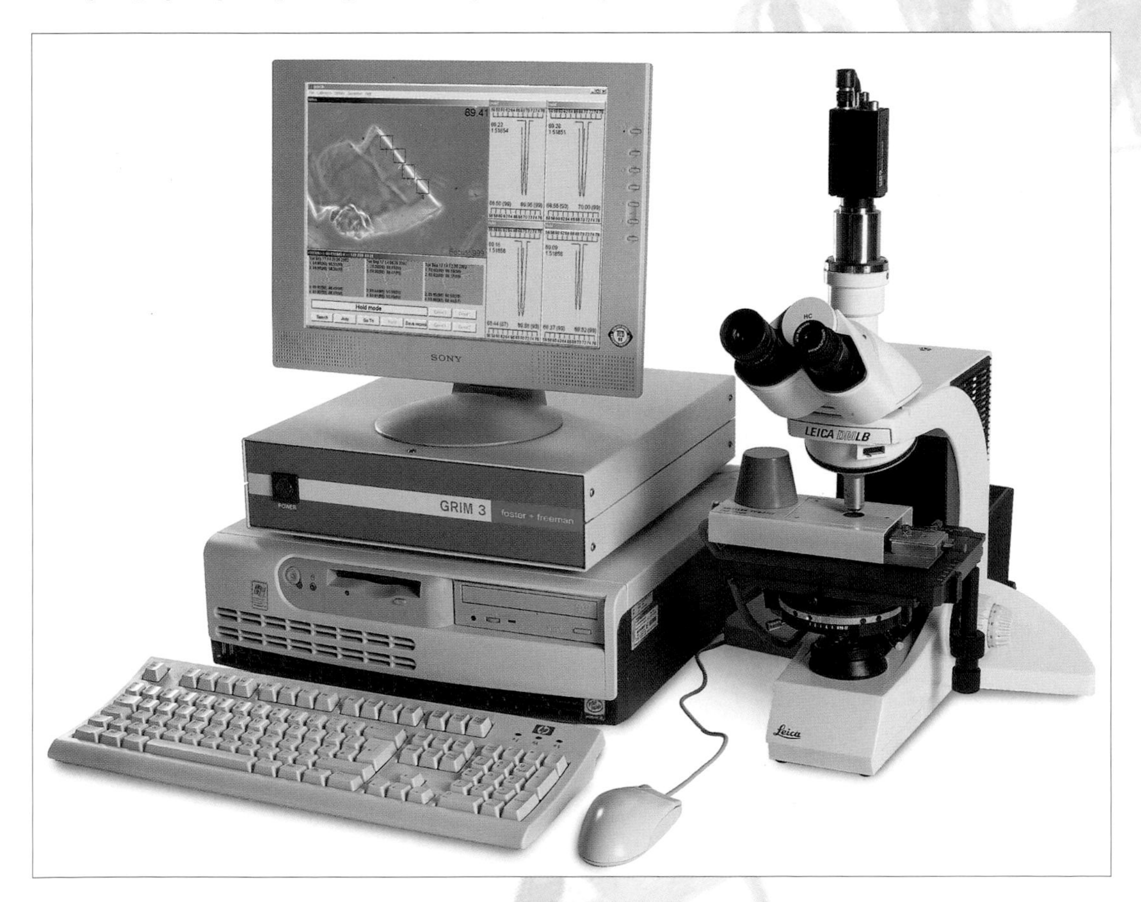

Forensics on the Screen

Forensics has become an increasingly popular topic for television and movies. Some of the best examples include:

- **Quincy, M.E.** (1976–83 TV drama series)—starring Jack Klugman in the title role as a relentless Los Angeles assistant medical examiner.

- **Manhunter** (feature film, 1986)—directed by Michael Mann. This is the forerunner to *Silence of the Lambs* and was remade as *Red Dragon* by director Brett Ratner in 2002. As a study in profiling, it's a classier production than the Hannibal Lecter films that followed.

- **Silence of the Lambs** (feature film, 1991)—directed by Jonathan Demme, based on the book by Thomas Harris, starring Jodie Foster and Anthony Hopkins. A young FBI agent seeks the help of serial killer Hannibal "the Cannibal" Lecter in pursuing another psychopathic murderer, on the basis that it takes one to know one.

- **Autopsy** (1994–, HBO documentary)—series hosted by famous forensic pathologist Dr. Michael Baden. Each program considers half a dozen or so cases, some historical.

- **CSI** (2000–, CBS TV drama series)—follows the adventures of a forensics team in Las Vegas, with parallel series set in Miami and New York. This is the big one, having passed the 100 episodes mark in 2005.

- **Minority Report** (feature film, 2002)—sci-fi thriller directed by Steven Spielberg and starring Tom Cruise, based on the Philip K. Dick story about forensics in Washington, D.C. in 2054. The ultimate triumph of forensics: the clues are discovered and used to convict the criminal *before the crime is committed.* But of course, like all forensics, the system is not infallible.

David Caruso in a scene from *CSI: Miami*. The appeal of the program and others like it is partly just the same as a Sherlock Holmes novel: following the clues. And the producers take the forensic content very seriously. There is a glossary of forensic terms on the official Web site. A further attraction is the implication that whatever horror the world throws at us, science can save us: the facts will out.

Index

Page references in *italic* refer to captions.

Acknowledgments

The Publishers would like to thank in particular Foster & Freeman, designers and manufacturers of scientific instruments for police and forensic science laboratories worldwide.

Anova Image Library: 47, 76, 81, 85, 122, 123, 140, 163, 166, 170, 179, 183.

Courtesy Nigel Blundell 102, 135.

Courtesy Michael Cherry, Cherry Biometrics: 158. http://www.cherrybiometrics.com.

CORBIS: 3 © PAUL HACKETT/Reuters/Corbis; 6, 8 © Reuters/CORBIS; 10 © Karen Kasmauski/CORBIS; 15 © Reuters/CORBIS; 24 © Richard T. Nowitz/CORBIS; 25t, 25b © Bettmann/CORBIS; 26, 27 © Reuters/CORBIS; 28 © CORBIS SYGMA; 30 © T.GERSON/L.A.DAILY NEWS/CORBIS SYGMA; 22 © CORBIS SYGMA; 33 © SINER JEFF/CORBIS SYGMA; 34 © Bettmann/CORBIS; 38 © Reuters/CORBIS; 40 © DAILY CAMERA/CORBIS SYGMA; 42 © KOESTER AXEL/CORBIS SYGMA; 49 © SIMON KWONG/Reuters/Corbis; 59 © Bettmann/CORBIS; 62 © Ralf-Finn Hestoft/CORBIS; 64 © Reuters/CORBIS; 70 © Bettmann/CORBIS; 71 © JOHN SOMMERS/Reuters/Corbis; 74 © Bettmann/CORBIS; 75 © Shepard Sherbell/CORBIS SABA; 78 © Bettmann/CORBIS; 82, 86, 88 © CORBIS; 89 © Najlah Feanny/CORBIS SABA; 90 © CORBIS SYGMA; 92 © JASON COHN/Reuters/Corbis; 94, 97, 104 © Bettmann/CORBIS; 107, 109, 110, 113 © Reuters/CORBIS; 114 © AYDELOTTE ROD/CORBIS SYGMA; 117, 118, 120, 121 © Reuters/CORBIS; 124 © Sean Adair/Reuters/CORBIS; 126 © Reuters/CORBIS; 127 © Andrew Lichtenstein/Corbis; 129, 131, 136, 143 © Bettmann/CORBIS; 149 © ADREES LATIF/Reuters/Corbis; 150 © Matthew Mcvay/CORBIS; 153, 155 © Bettmann/CORBIS; 160 © Fifty, NT/Corbis; 161, 162 © Reuters/CORBIS; 164 © CORBIS; 165 © Reuters/CORBIS; 169 © Kim Kulish/Corbis; 172 © Bruce Ely/The Oregonian/Corbis; 189 © Ted Soqui/Corbis.

Foster & Freeman 1, 9, 17, 44, 50, 51, 58, 60, 61, 106, 108, 112, 145, 187.

Courtesy Dr. Lowell J. Levine 7, 65, 98, 101, 133, 175.

Library of Congress Prints and Photographs Division Washington D.C. 20540: 13 [LC-B2-50-12], 14 [LC-USZ62-30516], 19 [LC-USZ62-90825], 20, 21 [LC-USZ62-105792], 22 [LC-USZ62-134663], 23 [LC-USZ62-116807], 36 [LC-USZ62-63765], 56 [LC-USF344-000878-ZB], 57 [LC-USZC2-1126], 66 [LC-B2-2645-11], 72 [LC-USZ62-77019], 73 [LC-USZ62-121399], 77 [LC-USZ62-133361], 80 [LC-USZ62-112609], 132 [LC-B2-2241-4], 134 [LC-USZ62-130710], 138 [LC-DIG-proke-20068], 139 [LC-USZC2-1457], 177 [LC-USZ62-133197].

Courtesy Mütter Museum, Philadelphia: 68.

Courtesy National Oceanic and Atmospheric Administration: 125.

Courtesy Sandia National Laboratories: 128

Science Photo Library: 53 TEK IMAGE/SCIENCE PHOTO LIBRARY; 55 GUSTO/SCIENCE PHOTO LIBRARY; 79 MICHAEL DONNE, UNIVERSITY OF MANCHESTER/SCIENCE PHOTO LIBRARY; 141 PETER MENZEL/SCIENCE PHOTO LIBRARY; 144 DAVID PARKER/SCIENCE PHOTO LIBRARY.